The Wealth of Humans

RYAN AVENT

The Wealth of Humans

*Work and its Absence in the
Twenty-First Century*

ALLEN LANE
an imprint of
PENGUIN BOOKS

ALLEN LANE

UK | USA | Canada | Ireland | Australia
India | New Zealand | South Africa

Allen Lane is part of the Penguin Random House group of companies
whose addresses can be found at global.penguinrandomhouse.com.

First published 2016
001

Copyright © Ryan Avent, 2016

The moral right of the author has been asserted

Set in 10.5/14 pt Sabon LT Std
Typeset by Jouve (UK), Milton Keynes
Printed in Great Britain by Clays Ltd, St Ives plc

A CIP catalogue record for this book is available from the British Library

ISBN: 978–0–241–20103–9

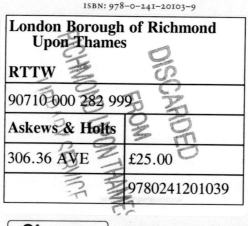
MIX
Paper from
responsible sources
FSC
www.fsc.org   FSC® C018179

Penguin Random House is committed to a
sustainable future for our business, our readers
and our planet. This book is made from Forest
Stewardship Council® certified paper.

A man must always live by his work, and his wages
must at least be sufficient to maintain him.

Adam Smith, *The Wealth of Nations*[1]

Don't mourn for me, friends, don't weep for me never,
For I'm going to do nothing for ever and ever.

Epitaph for a charwoman, traditional,
quoted in 'Economic Possibilities for our
Grandchildren', John Maynard Keynes, 1930[2]

Contents

Introduction

In January of 2014, *The Economist*, my employer, published a piece I had written on the future of work in an age of rapid automation. A sample:

> Ten years ago technologically minded economists pointed to driving cars in traffic as the sort of human accomplishment that computers were highly unlikely to master. Now Google cars are rolling round California driver-free no one doubts such mastery is possible ... A taxi driver will be a rarity in many places by the 2030s or 2040s ... bad news for journalists who rely on that most reliable source of local knowledge and prejudice.[1]

Not long after, a minor earthquake rattled the city of Los Angeles early in the morning. Within minutes, the first news report on the quake hit the wires:

> A shallow magnitude 4.7 earthquake was reported Monday morning five miles from Westwood, California, according to the US Geological Survey. The temblor occurred at 6:25 a.m. Pacific time at a depth of 5.0 miles.

What's notable about this second piece is not its content but its author, a piece of software ('Quakebot') developed by a programmer at the *Los Angeles Times*.[2]

The two pieces, mine and the robot's, were not especially alike. Quakebot is less given to chin stroking, for a start. My story was the product of months of research, reporting and writing: time spent building a view of the world and crafting an argument to support that view. It contained telltale signs of the author's attempts to make it interesting to readers. But they were both recognizably journalism:

intelligible, grammatical, informative. Journalism might outlast driving as a profession, but not, perhaps, by as many decades as we ink-stained scribblers would prefer.

Neither is automation the only threat to our livelihood presented by the digital revolution. At the time of writing, the inflation-adjusted value of advertising in American print newspapers has fallen back to a level last seen in 1950.[3] It may soon touch an all-time low. Let's be honest: it may soon touch zero.

The digital revolution is now teaching journalists and other workers of the rich world what a tectonic economic transformation feels like. It is putting us in the shoes of our great-great-grandparents: those who first experienced the transmission of a human voice across an electrical wire, who watched as the time to travel from one city to a distant other shrank from weeks to hours, and who found themselves displaced from jobs as smiths or farmhands by fantastic new technologies.

We have all found our working lives altered by it. Older workers might recall a time when factory work was still good work, easy to find, even for those without much education. Or they might remember a time when offices were jammed with clerical staff hammering at their typewriters and shuffling piles of paper around. But the pace of change is such that even the youngest members of the labour force can remember a different world. Services such as Uber and Airbnb, virtually unknown at the beginning of this decade, are fundamentally transforming industries that employ millions of people. Products such as Slack, a chat service designed to make it easier for colleagues to collaborate, are altering communication within workplaces, and clever bots that can email your contacts or order you lunch participate in the conversation just like human colleagues.

The pace of change particularly disorients workers in their forties or fifties, those whose decades of experience as a taxi driver or an administrative assistant might suddenly become less remunerative, or even worthless, in the years of work left to them before their planned retirements. And those now entering the labour force for the first time can have little confidence that their training will be of any use across the whole of their career – assuming that a career is a meaningful concept a half-century from now.

My own field has faced near constant disruption over the past couple of decades. Digital technology cost many printers their jobs long ago. Then came the internet, which allowed readers all over the world free access to a torrent of news and analysis, undermining subscription-based forms of journalism, while services such as Craigslist gutted newspapers' advertising revenue. Now firms such as Facebook and Apple are rolling out curated news feeds which promise to serve readers with the best stories from publications around the world – undercutting another of the valuable roles played by skilled editors. As a news consumer, this world thrills me; it is easier than ever to read brilliant journalism about all sorts of things, on subjects and from perspectives that might never before have got much of a platform. As someone who earns a living by the pen, however, I am nervous.

Our concerns are not simply about the uncertainty of employment in the years to come. Those of us who currently appear to have job security can more than likely look forward to making less in the future than we had once hoped we might. Over the last couple of decades, wages, adjusted for inflation, have scarcely grown throughout a broad range of rich countries – longer in some cases.[4] And this wage stagnation has occurred alongside other distressing trends. The share of income flowing to workers, as opposed to business and property owners, has fallen.[5] And, among workers, there has been a sharp rise in inequality, with the share of income going to those earning the highest incomes increasing in an astounding fashion.[6]

Wages have been rising in the fast-growing emerging economies, by contrast. But even there these other two trends – concentration of income in the hands of capital owners, and in the paycheques of the richest workers – are a growing source of concern.

Then there is the sobering data on employment. In America, the share of adult men of prime working age who are working or actively looking for work has fallen steadily, and in some cases dramatically, over the last generation. Among all men, the rate of participation in the workforce dropped from about 76 per cent in 1990 to 69 per cent in 2015.[7] That may not sound especially worrying, but it corresponds to a difference of about nine million men. And those squeezed out of work often find their lives upended. Stuck in atrophying communities with few prospects, many struggle to find purpose and satisfaction in life; indeed, recent research has turned up an alarming rise in mortality since the late 1990s among middle-aged white Americans, mostly accounted for by an increase in suicides and in drug and alcohol abuse. The authors see economic insecurity as a contributing factor.[8]

This trend is not limited to America, and neither can it be explained away as the product of ageing and retirement. In Europe, one in five adults under the age of twenty-five is unemployed.[9] Across the Organization for Economic Co-operation and Development (OECD), 12 per cent of people aged between fifteen and twenty-nine are neither in school nor work. Some are engaged in illicit activity or are in jail; others are in their parents' basements playing video games. Much the same is true of the long-term unemployed, many of them older men without much education, who drift around, often drinking to pass the day, lacking much, if any, connection to society at large.

For an awful lot of people, work has become a less certain and often less remunerative contributor to material security. It is a development that makes political forces of populist outsiders, such as Donald Trump and Marine Le Pen, and bestsellers of wonky economics books, such as Thomas Piketty's *Capital in the Twenty-First Century*,[10] an analysis of global inequality published in 2014 that flew off the shelves. Work is not just the means by which we obtain the resources needed to put food on the table. It is also a source of personal identity. It helps give structure to our days and our lives. It offers the possibility of personal fulfilment that comes from being of use to others, and it is a critical part of the glue that holds society together and smoothes its operation. Over the last generation, work has become ever less effective at performing these roles. That, in turn, has placed pressure on government services and budgets, contributing to a more poisonous and less generous politics. Meanwhile, the march of technological progress continues, adding to the strain.

THE CAUSES OF LABOUR ABUNDANCE

The digital revolution alters work in three ways. The first is through automation. New technologies are replacing certain workers, from clerks to welders, and will replace more in the future, from drivers to paralegals. Machines are becoming defter and software is becoming cleverer, and these improvements are increasing the set of human tasks that can be cheaply automated.

At the same time, the digital revolution has supercharged a second force: globalization. It would have been nearly impossible for rich Western firms to manage the sprawling global supply chains that wrapped around the world over the last twenty years without powerful information technology. And while China and other emerging markets might have become better integrated in the world economy even without companies such as Apple scattering production across the globe, such growth would have been much slower and less dramatic.

Instead, global employment grew by over one billion jobs over the last generation, with most of the growth occurring in emerging economies.[11] Workers there are, on the whole, less skilled than those in the rich world, and their incorporation into the global economy has been felt more keenly by workers in middle-skill manufacturing or back-office jobs than by white-collar professionals. That need not last; the developing world is home to millions of engineers, doctors, financial professionals and others who are just as capable of serving clients as their peers in America and Europe.

Thirdly, technology provides a massive boost to the productivity of some highly skilled workers, allowing them to do work which it might previously have taken many more people to accomplish. Technology enables small teams of money managers to run vast funds; it is increasingly allowing highly skilled instructors to build courses that can be taken and re-taken by millions of students, potentially replacing hundreds or even thousands of lecturers. New technology is allowing fewer doctors and nurses to observe and treat many more patients, fewer lawyers to pour through vastly more trial-related evidence, and fewer researchers to sift through massive amounts of data and test more hypotheses more quickly.

These three trends – automation, globalization and the rising productivity of a highly skilled few – are combining to generate an abundance of labour: a wealth of humans. In its struggle to digest this unprecedentedly enormous ocean of would-be workers, the global economy is misfiring in worrying ways. And the institution of work – apart from family, our most important piece of social infrastructure – can no longer be counted on to fulfil its many crucial roles – from the ordering of our days, to the allocation of purchasing power, to the strengthening of the social ties that are nurtured when individuals feel as though they are contributing positively to the community.

THE DIFFICULTY IN MANAGING A LABOUR GLUT

To say that humanity has too many workers is to defy a basic tenet of economics. Labour is not supposed to work like that.

When someone suggests that there are too many people around to do the work society needs done, he is said to be under the influence of the 'lump of labour' fallacy: the view that there is only so much work to go around – the lump. This view leads to policies such as those designed to lower the retirement age in order to create more work for the young. If we believe this basic theory, then we should certainly worry about the rise of machines.

Economists, however, are generally of the opinion that the economy works quite differently. They sometimes point to 'Say's Law', the work of eighteenth-century French economist Jean-Baptiste Say,[12] which is often summarized in the phrase 'supply creates its own demand'. Thus, when older workers stay on the job longer, they earn more money, and when they spend that money they create demand for other goods and services, leading to jobs supporting those goods and services. As far as labour-saving technological change goes, economists believe that when a person loses a job to a machine, it results in savings for someone – to the owner of a firm, or to consumers in the form of lower prices. This, in turn, leaves more money to be spent elsewhere, and that spending ought to create jobs for the displaced workers.

This magical reallocation is thought to occur because of the wonders of flexible prices and wages. An unemployed person looking for work is like a merchant selling a product. If the merchant cannot sell his wares, it means the price is too high, therefore he has two options: he can improve the product's quality, or he can reduce its price.

Think about a nineteenth-century craft producer of textiles: a moderately skilled worker who had been earning a decent living before the arrival of competing factories. Say that worker earned $3 a week as a self-employed craftsman. Then along comes a factory, which can churn out masses of cloth employing unskilled labour at $1.50 a week. The craftsman keeps on trying to sell his wares for a while, but then gives up. The mass-produced cloth is too cheap; he cannot sell enough at the higher price to sustain himself. Resigned, he wanders down to the factory and offers his labour services to the manager at $3 a week. The manager, of course, will chuckle at this offer and send the worker away. And the worker will tramp home disappointed, an unemployed victim of technology.

Maybe the worker then lazes around a bit, doing the nineteenth-century equivalent of watching daytime television, all while hoping the factory gets hit by a meteor. When he starts to run out of money, he visits other factories to see if they happen to be in need of some-one with his skills at a $3-a-week wage. But, to be counted as unemployed, a worker needs to be actively interested in finding work, and if the worker *is* in fact interested in finding work, then eventually he will realize what he must do. On his walks to various factories he will have noticed that a few were hiring engineers, at $5 a week, to maintain the equipment. He can therefore invest his time and resources in learning the skills to get a $5-a-week job, or he can accept $1.50 a week and find employment among the unskilled floor workers.

Economists don't believe in the lump problem – the idea that there are only a finite number of jobs in any economy. But they do acknowl-edge the severe disruption that comes to the individual worker when displaced by new technology. That person has two options: to learn to live on lower wages, or find a way to acquire more valuable skills.

Obviously, the ease with which these transitions are made very much depends on how many people are trying to make them at once. It is easier to retrain a few hundred workers than a few million. The hiring process takes time, and when the number of applications per job opening soars, employers can afford to be choosy. Eventually firms will come along that have thought up clever new ways to use this vast reservoir of under-employed workers, as cheap labour is a production opportunity, but that process can take a very long time.

And, all the while, the capacity to use technological solutions to do tasks for which humans have historically been relied upon grows. And grows.

The global labour force, which, as we have seen, grew by more than a billion workers over the last generation, will add close to another billion over the next. At the same time new technologies will make it ever easier to automate the simple work in factories, warehouses and shops that has historically accounted for a huge share of global employment. Technologies will also alter fields such as education and medicine, by allowing a few teachers or doctors to do work previously done by many.

The economy, and society, will try to adjust. That adjustment will mean stagnating wages for many workers, rising inequality, and a tenuous and fading connection to the world of work for many others. Workers are unlikely to take these woes lying down. Something has to give. Either society will find ways to shore up work or develop substitutes for it, or workers will use the political system to undermine the forces disrupting their world.

THE POLITICAL CHALLENGE OF
PROSPERITY

This *should* be a good problem for mankind to have. An abundance of labour is arguably the point, to the extent that there is one, of technological progress. It is the beginning of the end of the need to work hard to stay alive. A system in which people actively seek out labour they would strongly prefer not to do – manning call centres to handle the complaints of unhappy customers, or carrying packages around a boiling warehouse, for example – is not one society ought to aim to preserve any longer than technologically necessary. If society can find ways to automate such unpleasant tasks, or to share the work more broadly so that individual workers devote fewer of their waking hours to hard, unpleasant labour, that surely represents human progress.

For modern economies with more labour than they know what to do with, technological abundance creates the possibility of such progress. Like a massive gold mine or oil strike, powerful new digital technologies are a potential source of enormous wealth: one that can be realized without the need to keep everyone in society working. Utopia might, then, seem to be waiting just over the horizon (as a number of recent books, such as *Postcapitalism* by Paul Mason,[13] argue); all that must be managed is the slow reduction of hours devoted to menial work, combined with the distribution across society of the common wealth generated by productive technologies.

But is this better world of work achievable? Scholars have been imagining it for generations. In 1930, the British economist John Maynard Keynes wrote an essay describing his view of how the economic future would unfold.[14] At the time, the world was caught in a deepening depression. 'We are suffering just now from a bad attack of economic pessimism,' Keynes noted in the opening to his essay, 'Economic Possibilities for our Grandchildren'.

Yet in the piece he invited readers to look past short-term troubles to the remarkable long-run process of growth and progress in which humanity was engaged. After long millennia of labour in which living standards grew imperceptibly slowly, the societies of northwest Europe had, in the two or three centuries leading up to the depression, made a clear and extraordinary break with the economic past. Thanks mostly to technological progress, these societies had been enjoying phenomenal increases in wealth. And despite the woes of the Depression, Keynes rightly saw little sign that the underlying technological progress had ground to a halt.

Keynes believed that, once the world had overcome its Depression, growth would resume and living standards would return to the upward path they'd been on previously. He acknowledged that rapid technological improvements would cause some short-term discomfort ('a temporary phase of maladjustment'), but urged readers not to lose sight of the big picture:

All this means in the long run *that mankind is solving its economic problem*. I would predict that the standard of life in progressive countries one hundred years hence will be between four and eight times as high as it is today. There would be nothing surprising in this even in the light of our present knowledge. It would not be foolish to contemplate the possibility of a far greater progress still.[15]

Contemplating this progress, he concluded that it would free humans from concerns about the meeting of their basic needs. Time spent working would dwindle to perhaps fifteen hours a week, and then to nothing. And the main problem humanity would face would be just what to do with itself in a world of abundant leisure.

Keynes's forecast of progress in living standards has proven correct. Income per person, adjusted for living costs, has grown much as he foresaw; rich economies have already experienced at least a fourfold improvement in living standards.[16] It seems likely that at least some will, by 2030, have enjoyed an eightfold rise. Where, then, is the abundance? Where is the life of ease? Where are the fifteen-hour work weeks?

As it turned out, his description of humanity's economic problem was incomplete. Keynes worried that people would be bored in an era of technological prosperity; he didn't agonize over the possibility that politics would prevent it from ever arriving. As the years have passed and the global economy has continued to grow, it has become clear that the hardest part in finding utopia is not the figuring out of how to produce more. We've managed that. The hard part is the redistribution.

What we have not managed to do is to allocate the fruit of our production evenly enough to allow broad-based reductions in work hours. We haven't done that because it is politically a very hard thing to do. Crafting a balance of work and redistribution that is sustainable is *incredibly difficult*. The rich and privileged don't want to subsidize the poor. The poor may conclude that what redistribution the rich offer leaves an impossibly huge, even unfair gap in the incomes of the haves and have-nots. The poor may also not be content with an economy in which they are effectively unnecessary, kept at peace by a hand-out from the state. If redistribution is managed too clumsily, the incentive for clever or ambitious individuals to work to improve the economy might be lost, leading to stagnant growth and too little social surplus with which to provide all members of society with a rising standard of living.

Keynes should perhaps have foreseen the difficulty; in his day, he was a keen enough observer of the state of politics. By the 1930s, when the world was having its 'bad attack of economic pessimism', Europe had already gone through more than a century of bitter class conflict over the spoils of the industrial economy: 150 years in which the threat of worker unrest or revolution was a constant worry among the elite. Yet progress appeared to be on the side of the wage labourer. Time and again, workers asserted their power and won: the right to organize into labour unions, expansion of the franchise to men without property and (eventually) to women, establishment of labour-oriented and socialist parties. By the end of the Second World War, workers' victory over their employers seemed near absolute. A communist empire grew across eastern Europe and Asia, while, in the post-war West, the state also grew, managing large swathes of the economy, squeezing the rich with high rates of tax, and providing an ever more sprawling and generous 'cradle to grave' welfare state.

But political winds shifted. Communism proved a poor way to organize an economy. Technological progress and trade slowly chipped away at the power of organized labour. The prosperity of the post-war decades created a propertied middle class – increasingly well-educated and white collar – which over time grew ever less sympathetic to the priorities of the Labour left. In the 1960s, intellectuals like Milton Friedman[17] made an increasingly vocal case for a different, more market-oriented sort of economy. And, finally, the exhaustion of the unprecedented, glorious post-war economic boom and the arrival of the disappointing growth and high inflation of the 1970s created the conditions for a political break.

The break was more complete in some countries than in others. In the Anglo-Saxon economies – America, Australia, Britain, Canada and the like – the tax burden on the rich fell, the state liberalized and deregulated the economy, and the power of organized labour shrank dramatically. In Nordic economies, governments privatized and deregulated with gusto, but left in place a robust welfare state, and the taxation to support it. Continental economies, such as France and Germany, charted a middle course: liberalizing their economies and scaling back the welfare state in places, yet also leaving in place a considerably more interventionist and redistributive state than survived in America and Britain.

Most of us now of working age were born into a world in which this break had already begun. We inherited an idea of work that reflected this long struggle. It was a view of work as a positive good: economically necessary and morally beneficial. When work works, we understood, it provides a basis for a stable social order. It gives people something to do. It gives workers the sense that they are contributing to society and to the welfare of their families. It allocates income in a way that – if not always seen by everyone as just – is accepted by most as a valid basis for the distribution of resources. It encourages people to seek out the things at which they are comparatively good and to develop those skills. It makes the world 'go'.

Yet over the last generation it has become clear that history has not ended; that the political battle over the spoils of economic growth has not ended. While the liberalizing consensus of the last few decades formed and wrought its changes, the processes of technological progress and global economic integration transformed the economic role of the typical worker. Time and progress opened a gap between the prosperity society could potentially enjoy and the prosperity society, as currently structured, is capable of providing. A new political break looms.

Creating change in society to account for the abundance of labour will mean a resumption of the historical battle. It will be a battle between ideas – some new, some recovered from history's dustbin. It will be an individual struggle – what the hell should I do with my day? How and what do I teach my kids about a life well led? How do I provide for my family? And a societal one – how should we tax the fantastically rich? What does the state owe a middle class whose incomes have not grown for most of the last two decades? How welcoming should residents in advanced economies be to those who wish to move there from other countries in search of better lives, or to poor places that want to sell their goods and services to rich consumers? (And similarly, how passively should the world's poorer countries accept an isolationist, or nationalist, turn in richer countries?) If we can't offer our children meaning and identity in work, how do we channel their energies towards healthy alternatives, rather than ideological extremism, or social nihilism?

We are already seeing a rise in the appeal of extreme populist candidates blaming immigrant populations. In France, the anti-immigrant,

euro-sceptic National Front of Marine Le Pen is creeping dangerously close to the French presidency. Hungary's prime minister,
Viktor Orbán, has maintained popularity despite his authoritarian
tendencies by playing to Hungarian nationalism. And, in America,
Donald Trump has mounted an insurgent campaign for the Republican nomination for the presidency on a platform of virulent anti-
immigrant and anti-Muslim rhetoric. The nationalist right is ascendant around the rich world.

So, too, is a more radical left. This new left, however, has not yet
enjoyed as much electoral success as the radical right. The hard-left
Jeremy Corbyn shook the British establishment by taking control of
Britain's Labour party, but he has not been able to wrest control of
the government from the Tories. Bernie Sanders, a long-time socialist
senator from Vermont, mounted a surprisingly strong challenge to
Hillary Clinton for the Democratic nomination for the presidency,
yet ultimately fell short of the mark. Some radical left parties have
done somewhat better. The anti-austerity leftists of Greece's Syriza
party, for example, won control of parliament in early 2015 and
attempted to win a reprieve from the austerity policies imposed on
Greece by its European creditors (who were, in their defence, helping
to finance Greece's unaffordable debts).

Radical movements on both the left and the right are, for now,
relatively modest in their aims. The European right is pushing, in
some cases, for greater national sovereignty (or even an exit from the
European Union) and tighter controls on immigration. They are not
yet mounting a broad assault on liberalism and democracy – though
that may come. The left, meanwhile, is advocating an end to austerity policies in some cases and expansions to the welfare state in
others. Sanders campaigned on free college tuition and the creation
of a single-payer health insurance system. They are not yet running on confiscatory taxation and nationalization of the means of
production.

Both political extremes might never have the opportunity to pur-
sue their aims to their logical conclusion. But radicalism will become
an increasingly real and powerful force in global politics until governments begin answering the difficult questions posed by the digital
revolution. While people are dissatisfied and alienated, they will

continue to demand something better. A fierce contest of ideas and ideologies will follow, as radicals wrest control of the levers of power from conservative elites and put their ideas into action, for better or for worse.

As foreboding as this sounds, we can take some small comfort in knowing that we have been through all of this before. The industrial revolution destroyed old social orders in a similar way – wiping away whole swathes of employment, replacing workers with machines, widening inequality, and contributing to the marginalization of once-powerful political and social institutions. Radical new political movements then rose in response: labour unions; progressive social campaigns, which pushed for expanded suffrage, investment in education, temperance, and all sorts of other goals; and radical ideologies, such as anarchism, communism and fascism. The political and social contests among these groups led, ultimately, to new conceptions of the state and the role it ought to play in individuals' lives. Before the industrial revolution, the broad social role of the state we now take for granted – which includes universal education, publicly provided healthcare and financial support for the poor and out-of-work, generous pensions, the building and maintenance of networks of infrastructure – was unimaginable.

The worrying thing, of course, is that the world nearly ripped itself apart getting from point A to point B. From early in the nineteenth century to well into the twentieth, revolution was a constant threat in many rich countries. Governments struggled to tame financial and business cycles of increasing viciousness, which swept across advanced economies, destroying livelihoods and nest eggs. And nations fought bitter, unimaginably costly wars, culminating in the great ideological war that began in 1939 and claimed tens of millions of lives. That war, in turn, led to the development of weapons that threatened the very survival of humanity, and, it could be argued, did not truly end until the dissolution of the Soviet Union in 1991. The path to prosperity was a long and brutal one.

But, at the end of that road there was prosperity – for much of the world's population at least. A century ago, when the world was already more than a century into the industrial revolution, many of its benefits had yet to reach my great-grandfather, who toiled in poverty as a blacksmith in southern Virginia. America was the world's richest nation at the time (having surpassed Britain in income per person, adjusted for inflation, in the first decade of the twentieth century) yet much of the country still lacked electricity and running water, and many earned incomes not much different from those of workers in Medieval Europe.[18] I'm not sure my great-grandfather would have believed that, just eighty years later, his grandson and great-grandson would enjoy a standard of living that would have been the envy of ancient kings – and which was perfectly common among middle-class Americans of the late twentieth century – relaxing on a couch in front of a large colour television in an air-conditioned home with two cars in the garage, a full pantry, and a refrigerator stocked with cold drinks. Never before in history have so many people been so well off as at this moment in time.

But the next shoe is about to drop. Before we make it to point C – a world in which the benefits of the digital revolution are shared broadly and peacefully – we can expect difficulties. They have already begun.

The subject of the future of work in a digital economy has been well covered – in serious magazines, including but by no means limited to my employer, *The Economist*, and in a growing number of important books. Worries and speculation have grown more intense and more common since 2011, when Erik Brynjolfsson and Andrew McAfee published *Race Against the Machine*,[19] which laid out in compelling detail how quickly the capabilities of clever software and robots were improving. Authors like Martin Ford, whose 2015 book *Rise of the Robots*[20] described a vision of a post-work world, argue that robots and machine intelligence will create a world wholly different from anything that has come before, and that a techno-socialism of sorts will need to be adopted to keep society functioning. Economist Thomas Piketty's aforementioned masterpiece, *Capital in the Twenty-First Century*, set out a bold theory of inequality and predicted trouble ahead, as did Chris Hayes, whose book *Twilight of the Elites*[21] was an incisive examination of the loss of faith in elite institutions and technocrats, who have struggled to manage recent economic change.

Yet at the moment there is little agreement on how seriously to take automation concerns, how a transition to something like a jobless future might unfold and what ought to be done about it. Techno-optimists, such as venture capitalist Marc Andreessen,[22] lampoon the worriers as luddites and point to rising employment around the world as proof that their fears are overblown, while many left-leaning thinkers continue to blame globalization and the erosion of worker bargaining power, rather than robots, for stagnant pay and rising inequality in rich countries. Some writers, like Brynjolfsson and McAfee, and also Tyler Cowen, whose 2013 book, *Average is Over*,[23] speculates about America's economic future, anticipate a future in which broad economic and social change occurs incrementally, and in which sensible policy reforms (to education, for example) can make a technologically induced decline in the need for labour easier for households to manage.

The various partisans are like the allegorical blind men describing different parts of an elephant: each has his insights, but the competing stories have yet to be reconciled with each other. This book will provide that reconciliation. What is missing from the conversation is a clear explanation of how rapid technological change is compatible with both rising employment globally and disappointing growth in wages and productivity. And while it may be correct, as post-work prophets such as Ford foresee, that a world of technological prosperity and plenty awaits us in the distant future, it is wrong, I would assert, to characterize the digital revolution as something entirely different from anything that has come before.

On the contrary, as this book will argue, the digital revolution is very much like the industrial revolution. And the experience of the industrial revolution tells us that society must go through a period of wrenching political change before it can agree on a broadly acceptable social system for sharing the fruits of this new technological world. It is unfortunate, but those groups that benefit most from the changing economy tend not to willingly share their riches; social change occurs when losing groups find ways to wield social and political power, to demand a better share. The question we ought to be worried about now is not simply what policies need to be adopted to make life better in this technological future, but how to manage the fierce social battle, only just beginning, that will determine who gets what and by what mechanism.

THE MAKERS AND THE TAKERS

The battle lines of the great social upheaval are already being drawn. Their defining questions: who deserves credit for generating economic bounty, and who has the right to claim a share of that bounty once it has been generated.

Many of those earning top incomes, in individualist America in particular, believe they are the overtaxed 'makers' in society. High incomes, some of them suppose, are the reward for effort, innovation and job creation. In 2014, Gregory Mankiw, an economist and former chairman of George W. Bush's Council of Economic Advisers, wrote that '[T]he richest 1 per cent aren't motivated by an altruistic desire to advance the public good. But, in most cases, that is precisely their effect.'[24] To ask the rich to pay an outsized share of a country's tax is both unwise, because it diminishes incentives to create, and unfair, because it diverts resources towards an unproductive group of 'takers'. Rightly or wrongly, this argument is politically seductive; in 2012, Mitt Romney, the then Republican presidential nominee, became closely associated with his dismissal of America's '47 per cent'.* Speaking at a fundraiser, Romney stated:

> There are 47 per cent of the people who will vote for the president no matter what. All right, there are 47 per cent who are with him, who are dependent upon government, who believe that they are victims, who believe the government has a responsibility to care for them, who believe that they are entitled to healthcare, to food, to housing, to you-name-it . . . These are people who pay no income tax. Forty-seven per cent of Americans pay no income tax.[25]

It is easy to understand why the world's very rich feel they are treated unfairly. Where the wealthy of the late nineteenth century tended to

---

* Gregory Mankiw was an economic adviser to the Romney campaign.

'earn' their money from inherited landholdings and securities portfolios, today's rich are more likely to be self-made, and more likely to work longer hours than those at the bottom of the income ladder. Most of them worked hard to develop their skillsets, took risks to build a career, and devoted long hours to their jobs – without all of which they would not have come by their high incomes. And, indeed, capitalist societies rely on those rewards, to at least some extent, to encourage people to make those investments of time and effort, without which society as a whole would be poorer.

But just because a person's efforts generate a fortune does not mean that the fortune would not have been created had that person opted to work much less. Microsoft would not have existed without Bill Gates, and Bill Gates's enormous fortune would not be his had he not worked hard and applied his ingenuity to the building of that company. But if there had been no Bill Gates and no Microsoft, the world's personal computers would not have gone without a dominant operating system. Other companies would have filled that niche and other men, arguably, would have made that fortune.

That is not to say that individual effort doesn't matter; it matters tremendously. But the wealth generated by individual effort depends entirely on the society in which that effort is applied. Had Bill Gates been born in and remained in Somalia, he would not be a tech billionaire. Indeed, had a teenaged Bill Gates somehow been taken to Somalia and a teenaged Somali brought to America in his place, Gates would almost certainly be poorer today than the Somali. Somali society does not support an economy that can generate high incomes, while American society does.

A makers-and-takers conception of the world is one that neglects the social foundation on which wealth is built. We aren't merely divided into makers and takers. We are participants in societies, operating according to a broad social consensus. When that consensus breaks down, the wealth goes away. Society either agrees a way to share its riches that most members find acceptable, or the system fractures and the social wealth available to everyone shrinks.

THE RISE AND RISE OF SOCIAL WEALTH

Wealth has always been social. The long process of cultural development that eventually yielded the industrial revolution was in many ways the process by which humanity learned ever better ways of structuring society in order to foster the emergence of complex economic activity. Wealth creation in rich economies is nurtured by a complex system of legal institutions (such as property rights and the courts that uphold them), economic networks (such as fast and efficient transportation and access to scientific communities and capital markets) and culture (such as conceptions of the 'good life', respect for the law, and the status accorded to those who work hard and become rich). No individual can take credit for this system; it was built and is maintained by society.

The digital revolution is increasing the importance of social wealth in two key ways. Firstly, new technologies increase our potential productivity and output as a society; because we are capable of becoming richer thanks to digital technology, the economic return to economically important social institutions, such as a government capable of enforcing private property rights, is rising. The gap between the incomes of societies capable of supporting these institutions and those that cannot is growing. In 1980, Americans were thirty times richer than residents of the Central African Republic. In 2015, they were ninety times richer.[26] (In contrast, America was forty times richer than China in 1980, as its market reforms were just getting under way; it is now only four times richer.)

And secondly, the small-scale economic processes that generate new knowledge and turn it into profitable, welfare-enhancing activity are also becoming more social, and less individual, in nature. The value-generating pieces of successful companies were once satisfyingly tangible: consisting of buildings and machines, patents and people. That is ever less the case. Company cultures, which shape worker incentives and determine how a business reacts to changes in the marketplace, have become much more important in the digital age. Today, more than 80 per cent of the value of Standard & Poor's

500* firms is 'dark matter': the intangible secret sauce of success; the physical stuff companies own and their wage bill accounts for less than 20 per cent: a reversal of the pattern that prevailed in the 1970s.[27] A large proportion of that dark matter is an amorphous 'know-how': the culture, incentives and tacit knowledge that make a modern company tick.

The Economist is like that; our journalists gather information from all over the world, analyse it, and filter it through our editorial structures in order to generate pieces of journalism people want to buy. So is the (somewhat more profitable) Apple. Apple's phenomenal riches are built not just on the talent of its workforce, but on a particular internal culture and workflow, which prioritizes design and relentlessly improves on products until they are near-perfect: a culture that competitors find impossible to imitate. Successful companies, be they Goldman Sachs or BuzzFeed, evolve a way of gathering, processing and acting on information that is critical to their success, and which cannot easily be replicated. The value generated by a firm's culture, just like the value generated by networks of people within cities, or by a country's economic institutions, is social rather than individual. Culture is a set of beliefs and habits held in common by many people, and which only reveals its nature when it is held in common by many people. Orders given by one boss are not a culture; rather, a culture is made up of a common understanding of how daily business ought to be done.

The income made possible by companies' social structures can't easily be attributed to any one person or employee. But people do receive individual paycheques, and those who rise to top positions in successful companies get the biggest ones. It takes hard work to rise into such roles: to become a top editor at a successful publication or a managing director at a profitable bank. The people who occupy those positions logically and understandably draw the connection between the hard work needed to rise to such heights and the rewards they receive. But there is a difference between working hard to help your company generate more value and then pocketing a handsome

---

* Standard and Poor's 500 (or S & P 500) is an index based on the market capitalization of 500 large companies listed on American exchanges (with a combined market value of close to $20 trillion).

salary as a reward, on the one hand, and working hard to beat out others for top jobs at firms where the culture is key to success, on the other. The culture generates the wealth, and the culture consists of individual roles; working hard to beat out the competition to occupy a lucrative role within a value-generating culture is not the same as working hard to generate the value. One of the critical fights of the digital era will be over how to share social wealth.

As social wealth becomes more important, fights about who belongs *within* particular societies – and can therefore share in that social wealth – will also intensify. Over the last generation, firms have grown ever leaner, aggressively outsourcing work not related to their 'core competencies'. In a recent book on the phenomenon, David Weil, an economist at Boston University, writes that several decades ago a giant media firm like Time Warner might have directly employed massive amounts of labour, right down to the cable guy who hooks up your TV.[28] Now cable installers often work on a freelance basis, contracting for jobs through a cable-installation company, which in turn serves as a client to Time Warner. Arrangements like this move workers outside the 'society' of the large firm. They provide firms with a way to reduce the effective cost of workers, and to shift risk on to them: since these workers may become responsible for their own benefits, for instance, or be the first to suffer when a downturn strikes.

Membership battles – fights over who belongs – are more pronounced in cities, where high housing costs prevent people from moving into and enjoying the benefits of the most productive parts of a country. Google, for instance, has sought for years to add affordable housing for its employees on a part of its campus in Mountain View, California. Residents of Mountain View have waged a bitter campaign to prevent this, however, citing concerns about traffic and the need to protect local wildlife. That may sound perfectly reasonable; its effect, however, is to prioritize the welfare of existing residents by excluding future ones. Between 2012 and 2014, employers in the San Francisco Bay Area added nearly 400,000 jobs, while the local housing stock grew by fewer than 100,000 units.[29] Unsurprisingly, San Francisco housing prices rose by double-digit annual rates over that period.[30] That was brilliant news for local home-owners, who captured an outsized share of the fruits of the local tech boom, but high housing costs shut off the region, and its jobs, to new workers. Firms might consider moving elsewhere, but they can only do so at great cost, because of the social nature of innovation in the digital era. Houses might be cheaper in Topeka than in Silicon Valley, but Topeka is a poor substitute for Silicon Valley; it lacks the Bay Area culture that translates the germ of an idea in a Stanford dorm room into a billion-dollar tech start-up.

National borders create the starkest divide between the rich and the rest. No form of exclusion is as consequential. In America, a typical household of immigrants from the Philippines earns about $75,000 per year, or more than ten times what they'd earn in their home country.[31] There is no anti-poverty programme in the world as effective as access to American society – to its institutions and economy and opportunities. For now, despite brewing nativism and fears of terrorism, America remains relatively open by rich-world standards; in 2012 it accepted a net of five million migrants from abroad.[32] A good thing too; immigration dramatically boosts the incomes of the migrants themselves, but migrants also contribute in myriad ways to American wealth. They commit fewer crimes than natives and are disproportionately represented among entrepreneurs. But, across the rich world, the door to migrants is being pushed shut. In a time of economic and social anxiety, voters are choosing to limit access to their wealth-producing cultures.

SHRINKING CIRCLES OF AFFINITY

So these two kinds of conflict – between individuals and society, and between society's insiders and outsiders – create the fundamental tension presented by the digital revolution. To take full advantage of its promise, countries must become better at sharing social wealth. Yet the better countries become at sharing social wealth among members, the greater the pressure to shrink the circle of social membership.

The social battles of the industrial revolution era mostly focused on the proper role of the state. People organized and fought for a new social order; great new cities and factories arose; and crusading reformers and opportunistic politicians built new institutions in an attempt to round off the sharp edges of the brutal new industrial life. After a long and fitful social negotiation, most rich countries arrived at a social democratic model, in which the state to one degree or another helps to provide education, infrastructure, healthcare and social insurance to the old, poor and unemployed. The state also regulates industries and sets standards, and it enacts laws laying out how firms can and cannot treat their workers.

The digital revolution will reopen these discussions, but it will also force a new argument into the light that will define the generation to come: who belongs? Societies will face the need to define the community of people entitled to share in the common, social wealth made possible by marvellous new technologies. They will face choices, about which characteristics are grounds for inclusion, and what insiders must do to earn and keep their place.

This fight will be an especially difficult one because the nature of social redistribution must change. The industrial revolution was an all-hands-on-deck effort; there were roles for even the least skilled of workers: from cleaning horse manure off bustling city streets to moving parts around a massive factory. The social contract built during this age was one that protected the safety of workers, which made sure they were paid fairly for the critical work they did, which insured them against unexpected hardship, and which helped workers provide for themselves when they were too old or too young to contribute.

But the promise of the digital revolution is an end to work. The logical endpoint is an economy in which clever software and dexterous machines and abundant energy mean that human work is unnecessary. We are generations away from realizing that promise, just as societies in the early nineteenth century were generations away from achieving the mass industrial prosperity of the post-war decades. But the battle to create the institutions that will eventually support mass digital prosperity has begun. Creating mass digital prosperity is not about building institutions which ensure that all workers benefit from economic growth; it is about building institutions which provide for people who do not work because their work is not necessary to generate economic growth.

It's hard to contemplate how such institutions might work and prove sustainable. It's hard to imagine society deciding to provide rich lives for able-bodied adults, not because of anything they have done but because a rich livelihood is their right.

But we are not entirely without models for this sort of institution. One place to begin thinking through the problem is the family. Consider mine.

I grew up in a comfortable suburban house on the outskirts of Raleigh, North Carolina. As with all suburban houses in that part of the country, there was plenty of grass to mow in the summer and leaves to rake in the fall, and on Saturdays, between the morning cartoons and the afternoon goofing off, my three brothers and I were expected to handle basic landscaping chores. These chores never took longer than two hours, and would have taken considerably less if we'd worked as hard as we moaned. We got an allowance for our trouble, but we hated the work all the same. My father, easily the most assiduous worker I have ever met, rarely bothered to hide his frustration with our complaining and lack of work ethic. He had grown up on a farm in southern Virginia, doing the kind of work we kids had never known and will never understand: hard, manual work that needed to be done to keep the family eating: picking cotton, cutting tobacco, digging peanuts. In hindsight, he handled our apocalyptic moaning about being asked to put a few acorns in a bucket with more grace and aplomb than we deserved.

Dad could have hired someone to mow the lawn, and his refusal to do so wasn't just a matter of money. Tending the lawn was about the *lessons* he needed us to learn: that while we would have plenty of time to play, our Saturday could not be entirely without structure. That while our parents might provide us with everything we needed, we should not take their generosity for granted, or conclude that it was right to enjoy such things without some effort to contribute to the family. Picking up acorns wasn't a matter of material necessity; we were fortunate in that our childhood labour never was. Instead it was an investment in the mutual goodwill that helps keep any society, including a family, functioning smoothly.

It has proven a valuable lesson.

An economy is not a family. But my father's weekly struggle to get his children to take just a little time out of their weekend to mow the grass or clear the lawn of acorns is not a bad way to understand recent troubles in global labour markets by comparison. Economic models can take us a long way in parsing what is happening to programmers in Seattle and textile workers in Dhaka – supply and demand and the productivity of labour matter – but changes in the nature of work, in what it means to have a job and in what one takes home at the end of the day for doing it, depend heavily on the *social context* within which the work takes place. As children, the incentives and the sense of purpose to our work were inextricably linked to the context within which we were working; the chores were not simply an economic transaction but a way for my parents to order our day, to impress upon us particular values, and to satisfy themselves that they were raising us well.

Work, done by adults in the global marketplace, is not all that different. The mission of this book is to explain why: to examine the challenge of ordering our lives and our labour in a world of technological abundance.

In the pages that follow I'll break the problem down into four main parts. I'll look first at what's happening on the ground as technological progress leads to accelerating social change and erodes confidence in the foundational institutions of industrial economies, from companies to global trade agreements. I'll then explore the key forces – economic, social and political – shaping the evolution of this new world of too many workers. I'll then zero in on the ways in which the abundance of labour is altering the operation of our economy – our cities, our financial markets and our trading patterns – in worrying ways. I'll conclude with thoughts about how we are likely to try to manage the change, and where we can expect to have most and least success.

I

The Digital Revolution and the Abundance of Labour

I

The General-Purpose Technology

Technological progress used to be something you could feel in your bones. It was the thing that was all around you, turning your world on its head. It was the sensation a young man might have felt when the arrival of mechanical harvesters made his labour on a farm in the countryside unnecessary, leading him to leave for the city, where giant steel-framed towers stretched upwards in what must have seemed like the very realization of the Tower of Babel, and where a rich man might occasionally zoom by in a wheeled vehicle that, astonishingly, powered itself along without the aid of horses. It was the end of an ancient way of doing things and its replacement with something entirely different and unknown.

The industrializing economies of the nineteenth century staged extravagant World's Fairs to celebrate the world's new wonders. These extraordinary gatherings, such as London's Great Exhibition of 1851 or Chicago's World's Columbian Exposition in 1893, look in hindsight like magnificent compressions of historical time: centuries of pre-industrial life crashing at high velocity into the modern world. And so in London, Queen Victoria, whose relations would sit atop many of Europe's centuries-old monarchies, opened the London exhibition, which featured working textile machinery, early photographic technology and one of the first examples of indoor flushing public toilets. On a visit to the Crystal Palace, where the exhibition was staged, the English novelist Charlotte Brontë gushed, 'It seems as if only magic could have gathered this mass of wealth from all the ends of the earth – as if none but supernatural hands could have arranged it thus, with such a blaze and contrast of colours and marvellous power of effect.'[1]

And in Chicago, William Cody's Wild West show was denied permission to operate within the fair itself and so set up, profitably, just outside. 'Buffalo Bill', as he was more commonly known, entertained visitors with visions of a rapidly vanishing frontier, itself a recent imposition on societies thousands of years old. At the nearby White City, among the many grand buildings built especially for the fair, the public was dazzled by electrical displays, from the lighting of the exposition itself to the wizardry of Nikola Tesla, an inventor and engineer who helped tame electrical current and develop electric motors (among other things).[2]

People came to these fairs to see wonders – and the nineteenth and early twentieth centuries had plenty of them. But people hardly needed to go to one of these expositions to know that great and powerful change was afoot. In 1840 Chicago was a speck on the map, with a population of less than 5,000. By the time of the Columbia Exposition fifty years later, it was America's second largest city, with more than a million people, and skyscrapers beginning to reach into the air above Lake Michigan.[3] Chicago's extraordinary rise was bound up with the arrival of the railroad, which transformed travel across the continent. Before the construction of the railroad, the stagecoach journey from New York might have taken a full month; in addition to the bumps, passengers faced the risk of breakdowns, accidents and general isolation along the long and lonely route. The arrival of the railways shrank the time needed to travel to about a day, changing the journey from a once-in-a-lifetime adventure to a commonplace.

And thanks to telegraphy, news, which had previously travelled at the same plodding pace as people and freight, now flew along at the speed of electricity: Chicagoans learned that they had been awarded the World's Fair at roughly the same time New Yorkers did. In the space of a lifetime, the world shrank from a place in which those living on the other side of the earth might just as well have been on the moon to one in which vast distances could be travelled in days, and people around the world lived and experienced the same news at more or less the same time. There was a dizzying, tangible acceleration in life that altered the world and the way people thought about it.

Life over the last sixty years has been quite placid, by contrast. The changes we have experienced are overwhelmingly of the incremental sort: televisions have become bigger, better and cheaper; automobiles are safer and more environmentally friendly, and have added bells and whistles, such as power locks and rear-window defrost. Life-spans have risen, but humanity didn't reinvent germ theory. Air travel became more ubiquitous, but we didn't reinvent powered flight. Dramatic, wrenching technological transformations occurred in a handful of economies: South Korea and Singapore, for example, and more recently China. But these were merely examples of the delayed arrival of the whirlwind that had upended rich countries in the nine-teenth and early twentieth centuries.

After so long a period of modest economic evolution, many of us have forgotten that economic advance ever occurs at any other speed. Some techno-pessimists, such as Robert Gordon, an economist at Northwestern University, argue that the slowdown is irreversible. Technological progress, he argues, gathered momentum over a long period of time thanks to a series of fundamental intellectual insights. The development of a deep understanding of what electricity is and how it might be used is not something that can easily – or perhaps ever – be duplicated. The inventions that followed on from advances in the science of electricity, and in other areas, are not like water drawn from a river but like coal mined from the earth: society couldn't help but exploit the most accessible, most abundant veins first, leaving only the marginal, difficult things for later generations (like ours).[4]

Worse, the pessimistic view runs, the deceleration in intellectual progress is itself evidence that there are few, if any, fundamental insights such as that into the science of electricity still remaining out there, waiting to be discovered. Humanity is far cleverer now than it was in the nineteenth century, they argue, and there are many more highly trained scientists and engineers working with vastly greater research and development resources. If there were an electricity-like breakthrough lurking out there in the shadows, humanity would have uncovered it already.[5]

Pessimists point to a parallel to this intellectual counsel of despair in the economy itself. Technological progress peaked during a period from the late nineteenth century to the mid twentieth century, they assert, an era sometimes called the 'second industrial revolution' (the first having been the initial factory boom in Britain, built on the taming of steam power). This second revolution wrought fundamental changes in the world: fantastic, one-off transformations that can't be repeated. It was during this period that rich economies became electrified. This was the era in which modern sanitation and indoor plumbing were developed, and in which cities grew to truly modern size, in scale and population. It was the period that gave us what are still today the most advanced personal mobility technologies: the automobile and the airplane. It was this period that made the modern world what it is.

It was also the era in which the modern job evolved: shaped by the rise of the factory economy, by unionization and the political mobilization of the working class, and by the construction of a social safety net. By its end, the second industrial revolution handed to society a template for modern life – one or two forty-hour-per-week jobs per household, supporting a consumption-oriented middle-class lifestyle – which has been the social foundation for rich economies for most of the last half-century.

For a new technology to be as powerful as the old ones it would need to create in the world something similarly transformative. It would need to create for humanity a life as different from reality today as the life of the 1960s was from that of the late nineteenth century. Alternatively, the pessimistic view of the arc of technology, up and down the sides of one great wave of advance, implies a pace of social change that is incremental. It implies a continuation of the pattern of the second half of the twentieth century, when the children of baby boomers could expect to do as their parents had done – go to college, get a good job, have a family and buy stuff, before retiring.

The pessimistic view is ever harder to square with the evidence of change all around us. It seems increasingly clear that the decades after the second industrial revolution did not represent a slide towards stasis but a lull in the process of headlong advance. The lull has been long enough to allow us all to forget what headlong advance feels like. The digital revolution will remind us. It has slowly grown in its transformative power over the last few decades, to the point at which it is increasingly capable of inducing the same sort of historical vertigo our ancestors experienced in the 1900s. There is no telling whether, when all is said and done, the digital revolution will prove as dramatic as the technological shifts of the industrial revolutions. But it will be dramatic enough: once again, the kind of change you can feel in your bones.

For the gathering pace of change, we can thank ever-more-clever computers, which are finding uses in ever more corners of the economy. There will soon be thinking machines everywhere.

REVOLUTION MACHINES

Computing is not simply another valuable invention, on a par with the washing machine or the photocopier. Digital computers represent something more fundamental: something powerful, which allows us to do things differently and better across all facets of life. Its proper analogues are steam and electricity.

In 1876 the first great exhibition to be held in America opened in Philadelphia: a Centennial Fair, part of the country's celebration of 100 years of independence. Britons were invited; a few showed off their newly developed penny-farthing bicycle – the one with the giant front wheel. But among the most impressive exhibits on display was the Corliss steam engine: a behemoth of a mechanical device, seventy feet high and weighing 650 tonnes. The 1,400 horsepower Corliss engine drove a system of belts that powered the whole of the fair's machinery hall.

George Corliss, an American engineer, patented his engine in 1849, more than eighty years after James Watt made his most critical contributions to steam-engine design. At the time, American manufacturers used a total of less than 2 million horsepower (or roughly the output of a large turbine in a modern power plant), most of which was generated by water. A half-century later, American manufacturers used more than 10 million horsepower in operating their factories, the vast majority of which was generated by steam engines, and the American economy was overtaking Britain as the world's leading industrial and technological power.[6]

Economic historians label things such as steam power as a 'general purpose technology': an advance that can be used to do things more effectively across many different facets of life. A steam engine could be hooked up to any production facility that previously relied on wind or water or animal power. It could be affixed to transport devices – boats, cars, train engines – to make them go farther, faster, with more horsepower. Steam could be used to boost productivity in all sorts of contexts and industries. It is the general-purpose technologies – such as steam and electricity – that generate economic revolutions. And computing is a fantastically powerful general-purpose technology.

Engineers tinkered with computing machines for millennia, but the pace of advance in mechanical computing truly picked up in the nineteenth century. Early computing innovation found its way into a loom invented by a Frenchman called Joseph Marie Jacquard, which used punch cards to 'programme' the loom to produce particular patterns in the fabric. In the early twentieth century, the vacuum tube (a light-bulb-like device in which an electrical current is transmitted from one electrode to another) became the guts of early electronic computers. Early computer scientists learned that the tubes could be used as electrical switches, which meant that they could be used to calculate.*

It was the Second World War, however, which transformed the computing world. Governments poured massive resources into the development of new machines that could be used to break codes or to model nuclear explosions, in the process laying the groundwork for the post-war computing industry. In the post-war years, engineers enjoyed great success developing critical new components (such as the transistors that replaced vacuum tubes), making them ever more powerful and shrinking them down. Smaller, cheaper and more powerful components gave rise to an enormous new personal electronics

---

* Switches, which can be set to one of two positions ('on' or 'off') can be used in binary mathematical operations, in which all figures are represented in combinations of ones and zeroes. The larger the number and faster the operation of the switches, the more powerful the computer. The earliest electronic computers typically had a few thousand switches. Microprocessors today tuck a few billion into a much smaller package.

industry, producing stereos, televisions, calculators, video gaming systems – and then personal computers and mobile phones.

Progress in computing owes much to 'Moore's Law'. In 1965 Gordon Moore, a co-founder of Intel, reckoned his industry could double the number of transistors in an integrated circuit roughly once every two years, and that this doubling would likely continue.[7] This astonishing pace of progress has been maintained for most of the last half-century, changing computing from something done at great expense by house-sized machines to something done all the time in tiny devices which now rest in the pockets of about 30 per cent of the world's population.

This slice of history played out during a period that economist Tyler Cowen, of George Mason University, has labelled the 'Great Stagnation'.[8] A half-century of extraordinary gains in computing power somehow did not return humanity to the days of dizzying economic and social change of the nineteenth century. In 1987 the Nobel Prize-winning economist Robert Solow mused, in a piece pooh-poohing the prospect of a looming technological transformation, that the evidence for the revolutionary power of computers simply wasn't there. 'You can see the computer age everywhere but in the productivity statistics', he reckoned, and he had a point.[9] Productivity perked up in the 1990s but wheezed out again in the 2000s.

And that, some seemed to conclude, was all there was. In the 2000s Robert Gordon began posing a thought experiment to his audiences: would they, he wondered, prefer a world with all the available technology up to 2000, or one with all available technology up to the present day *except for* indoor plumbing? His little test effectively made the point that what occurred in the second industrial revolution was powerfully transformative, in a way the advances of the internet age simply weren't. Google is grand, but it's not as transformative as running hot water.

What I like about this thought experiment, however, is that it unintentionally also makes the contrary argument. When Gordon began posing this question in his papers, the answer was so clearly the option *with* indoor plumbing as to make the question something of a joke – which is what Gordon intended. But with each year that passes, the choice becomes less clear. For many people in developing economies, a smartphone is *obviously* more important than indoor plumbing: the latter is nice, but the former provides an invaluable economic and social link to the global economy. Meanwhile, in rich countries, smartphone culture is now so deeply entrenched that people might (might!), if forced to make the choice, give up their toilet in order to keep hold of their phone. Nor are smartphones the beginning and end to the contributions of the digital revolution; amputees in possession of thought-controlled prosthetic arms could explain to Gordon that recent advances go well beyond social networks and dating apps.

The transformative capacity of the digital revolution has grown, steadily and surely, over the last half-century. Machines can now drive cars and carry on a basic customer-service conversation. They can spot faces in a crowd and provide instant, serviceable foreign-language translation. They can write reports and edit genomes. And machines that are powerful enough to do those things can do much more besides. Computing is beginning to make good on its promise as a general-purpose technology.

THE DIGITALLY DISAPPOINTING ERA

So what has taken so bloody long? The pronounced lag between the emergence of widespread computing and the beginning of revolutionary economic and social change can be blamed on two factors. Firstly, a remarkable new invention can't transform society until society has learned how to use it effectively. As Gordon himself notes, the productivity burst from the key innovations of the late nineteenth century played out over the course of the entire first half of the twentieth century. The key discoveries in the taming of electricity were made in the 1870s and 1880s, yet it was not until the 1920s that electricity was widely used in factories and households. Even after the promise of a new technology is apparent to all (or nearly all), it can take decades for society to reshape itself in order to reap the benefits of that technology.

Consider the automobile. Cars were an impressive piece of technology in the late nineteenth century, and their transformative potential was apparent by the early twentieth. Yet it took a very long time for societies to fully exploit that potential. Social and cultural norms needed to evolve; some of those norms then needed to be codified into a body of regulatory law: describing where people could drive and how fast, who was allowed to operate a vehicle, what repercussions there would be for misuse, and so on. The physical structure of society changed in response to the automobile. Governments spent vast sums to construct networks of streets and highways, while suburbs oriented around cars covered the landscape outside central cities. Firms experimented with car-oriented business models before coming up with hits along the lines of pizza delivery and NASCAR. And not until the last few decades of the twentieth century did the perfection of container shipping, trucking and big-box retail converge to transform the consumer experience in rich economies, as well as the development opportunities in the emerging economies that became the source for many of the cheap goods stocking Wal-Mart shelves.

The wholesale change in society that occurred alongside industrialization – including mass urbanization, a significant increase in the educational attainment of the population, great change in the size and role of the state *and* in how governments are chosen – were not simply knock-on effects of technological change but were ways in which society evolved in order to *enable* the productivity possibilities of the new technologies. To make good use of discoveries in industrial chemistry societies needed to educate plenty of chemists and chemical engineers, for example. That, in turn, required the development of robust primary and secondary school systems and the development of technical universities. Achieving *that* took the mobilization of pro-education pressure groups, the election of politicians sympathetic to the cause, investment in the schools themselves, development of curricula, and finally the education of cohorts of students. The re-forging of society is not a rapid process. For that reason, the full exploitation of the possibilities presented by a new technology takes a very long while.

Unsurprisingly, computing has faced its own adjustment period.

Impressive gains in processor speeds and the tumbling cost of memory do not themselves boost productivity. For that to happen, computer manufacturers must figure out how they can most attractively package computing components into devices that firms and households might want. Should they build and market mainframe-linked terminals? Or PCs that connect across an internal network? Programmers must figure out what problems those machines can usefully solve and write code to allow them do so. Firms must then sort out what combinations of hardware and software will help them save money or boost output. Should their employees use PCs or Apple machines? What database software should the firm run? Should all employees have a mobile phone? What kind?

When the firms think they've sorted those issues out, they need to buy the equipment, hire people with the skills to use it, and rearrange the way they operate to take full advantage of the new machinery. Students trying to figure out what to study at university must discover that firms are interested in people with particular computer-complementary skillsets and change their education plans accordingly.

Meanwhile, alongside the old businesses attempting to use new technology to make their existing practices more efficient, brand new businesses pop up and try to use newly available technology to try radical new approaches to old problems. While some legacy retailers adopt bar codes and software that can track inventory and keep tabs on consumer purchases, Jeff Bezos founds Amazon. As both sorts of firms experiment with new approaches, complementary businesses form or evolve in anticipation of retailers' needs: logistics businesses focused on warehousing and freight, or product sellers keen to take advantage of online marketplaces.

These repeated cycles of experimentation with new technologies, and of adaptation among firms, workers and consumers, generate the lag between the appearance of an innovation and observed gains in productivity or striking changes in lifestyles. Studies of information-technology adaptation reckon there is generally a gap of between five and fifteen years between investments in new technology and the appearance of measurable gains in productivity associated with that investment.[10]

When people survey the technology all around them, at work and in their homes, they are seeing the world of technology in a rear-view mirror. America's productivity boom of the 1990s is associated in popular memory with the crowds of consumer-facing dotcom businesses, such as Pets.com, which spent lavishly on Super Bowl advertising and existed mostly to give their founders a million-dollar payday when the firm went public. But what actually drove the rapid productivity growth of the boom were older and more prosaic technologies, such as the 'enterprise software' products sold by Oracle and SAP. Firms enjoyed massive productivity gains by using computers to keep track of their inventories and customer information, and from using that data to eliminate waste (by ordering new supplies on an as-needed basis, for instance, rather than keeping lots of extra inventory around just in case).

The survivors of the dotcom mania, such as Amazon and Google, made their most significant impact on society well into the 2000s. And many of the big technology stories of the business world right now have app-based companies at their heart, nearly a decade after Apple released its first iPhone and launched the 'app store'. The new businesses and hot consumer trends changing the world at any given moment are built on old technology. If humanity has underestimated the transformative potential of the digital revolution, that is partly because the consequences of a new advance often show up quite a while after the advance itself.

But that lag is not the only reason the digital era has taken so bloody long to wow us. In fact, some full-throated techno-optimists argue, information technology simply hasn't been that impressive for most of the last half-century. Yet that, they say, should in no way convince us that future progress will be similarly disappointing. On the contrary, a long period of modest progress is precisely what we would expect to see from a technology improving in exponential fashion from a very modest starting point.

In an influential 2012 book, *Race Against the Machine*, two MIT scholars of technology and business, Erik Brynjolfsson and Andrew McAfee, argue that people aren't very good at assessing the pace of exponential technological progress (for example, the repeated doubling in microchip power described by Moore's law).[11] They borrow a parable popularized by the futurist Ray Kurzweil.[12] In the legend, a wise man invents the game of chess and presents it to his king. Pleased, the king allows the man to name his reward. The wise man responds that he wishes only modest compensation, following a simple rule. He would have one grain of rice on the first square of the chessboard, two on the second, four on the third, and so on, doubling each time for each of the sixty-four squares. The king chuckles at the apparent measliness of these amounts and says yes. It soon becomes clear that he has made quite a big mistake. After two rows the king owes nearly 33,000 grains of rice and is not chuckling quite so much. By the last square of the first half of the chessboard the amount involved is enormous, totalling more than 2 billion grains, or nearly 100,000 kg, of rice – but it is not yet absurd. Yet on the first square of the second half the king must pay that entire sum again, and then twice that, until he owes a Mount-Everest-sized pile of rice.

The tale is meant to illustrate the deceptive nature of exponential growth. Decades of progress can yield meaningfully large improvements that nonetheless fall short of transformative change. But each generation of progress is as significant as the sum of all those that came before. Around the time that the process of advance reaches the first square of the second half of the chessboard, the capacities of cutting-edge technologies become truly breathtaking: machines can suddenly drive cars, or hear and understand human speech, or look at a photograph and describe exactly what they see – advances that looked unattainable just a few years before. Those advances open up dramatic and slightly frightening new economic opportunities. And just as the very first start-ups experimenting with the very first business models based on those technologies venture into the marketplace, the next generation of technological advance lands, and adds as much new power as the industry managed to develop in every previous generation of innovation – including the one before, which brought all that scary new machine capacity.

One can conclude too much from this narrative of progress, however. Processing power is not productivity growth, and cheap supercomputers in our pockets will not be economically transformative if we can't come up with economically transformative things to do with them. But it would be surprising if exponential advance in computing didn't generate dramatic economic change, given the general-purpose nature of the technology. Most of what humans do when they are working boils down to computing. Sceptics regarding the possibility of instant machine translation didn't argue that it was impossible because language was about more than computing; they argued that it was impossible because it required *really hard* computing. But really hard computing is precisely where exponential advance in information processing comes in handy.

If driverless vehicles were all the revolution managed to produce, the economic and social impact would be stunning. About five million Americans work providing 'transportation services', including about half a million cab drivers and nearly one and a half million drivers of freight trucks.[13] Autonomous vehicles could eliminate all of that work. But that would only be the beginning. Driverless vehicles might double as nannies, picking up youngsters from school and delivering them to a parent's office or an after-school activity. They could facilitate the near-complete automation of massive amounts of retail; many grocery shops might vanish as consumers could instead get into the habit of mentioning to their smartphone when a bottle of wine is needed, which could then be ferried from a nearby warehouse by autonomous car. Car ownership might itself become obsolete, since vehicles of any sort could be hailed instantly. Traffic might vanish in the space of a few years, while the massive tracts of land given over to parking lots could suddenly be used more productively.

But a computer that can operate a car effectively represents a technological capacity that can be applied in many, many other powerful ways: from machines that can sift through data to spot potentially worrying health developments to machines that not only do your taxes for you but talk you through worries about your business's sales strategy or concerns about your retirement plans. A capability threshold has been crossed. And while humans sort out how to exploit new machine capabilities to their fullest, machines are being made more capable still. The main protection human workers now have against machines is that the machines are not very smart; they write dry, boring news stories, for instance. But that is no protection; machines are much better at becoming smarter than people are.

THE PARADOX OF POTENTIAL

A dose of perspective is in order. It is important to remember that major technological revolutions usually generate enormous benefits alongside the disruption they cause. Higher productivity levels mean that firms can afford to pay higher incomes. Just as important, the march of technological progress has lengthened, improved and enriched our lives. Indoor plumbing helped to make cities tolerable, non-deadly places to live. Assembly-line techniques dramatically reduced the cost of goods such as cars and televisions, in the process turning them into basic consumer goods rather than the playthings of the very wealthy. Electrification upended all sorts of industrial processes, and also gave us electric light, telephone calls and rock music.

The digital revolution is no exception to this pattern. The web causes hardship for publishers precisely because it is so good for news consumers, who now enjoy access to massive amounts of information at very low cost. The global supply chains enabled by information technology have been hard on some workers but very good for shoppers as a whole, who now enjoy cheaper electronics, clothing and toys as a result. One marvels at the pure, massive consumer surplus generated by something like Wikipedia. When I was a kid, there were still people who would knock on your door to try to sell you encyclopedias, and school essays often needed to be written in a library, where you could easily turn to the Britannica on the shelves or dig through the card catalogue, looking for just the right source. Now anyone can dig through the free pages of Wikipedia: a source far more exhaustive (if not always *exactly* right, but then we needn't assume the books in the library were either), far more easily navigable, and available instantly in many languages; the mind boggles.

It is often these seemingly minor things that generate the most utility; the ability to Instagram or to video chat with parents on the other side of the Atlantic is an invaluable improvement on the communication options available a generation or two ago. The digital revolution also allows consumers to extract more value from old stuff: a dusty old text in a New England bookshop may be just the thing a reader in Omaha wants, and, thanks to online used-book listings, she now stands a good chance of finding it. New apps allow consumers to make better use of apartments or seats in their cars that would otherwise sit empty.

High-quality online courses could lead to massive layoffs of lecturers, but would make a good education easily affordable and accessible to people all over the world, people of all incomes and from all walks of life. Cheap wearable computers and computer monitoring and diagnosis could mean big trouble for lots of doctors and nurses, but should improve health while also reducing health-care costs. Driverless cars will displace professional drivers, but should save hundreds of thousands of lives thanks to reduced accidents.

The digital revolution is an irresistible force because it offers humanity so many good things. It forces society to face the trade-off: new and improved goods, services and experiences at lower costs in exchange for social and economic disruption. Labour-market woes are growing because humanity is choosing, decisively, in favour of the fruits of the digital age. We choose all the time: when we hail a car using Uber, when we buy a cheap smartphone assembled on the other side of the world, when we stop paying for cable television because we can stream everything we want to watch, when we rate plumbers on Yelp, when we book a holiday villa on Airbnb.

As technology improves, we will find ourselves lured into more fundamental changes. Going carless, or skipping a high-priced university in favour of online courses, will cease to be sacrifices forced on people by a lack of resources and will become the easier, more liberating decisions. We plunge into the unknown future because the technologies that transport us there offer us the promise of something better. For this reason, it is no good to wish technology away. It would be futile, and indeed immoral, in many cases to deprive people of the ability to improve their lives by exploiting advances in technology.

But we are not just consumers. Our ability to consume depends upon our ability to produce. While falling prices amid expanding choice may eventually spread to housing and healthcare, food and energy, technology has not yet enabled the possibility of a utopia in which all necessities can be had for the asking. We still need the purchasing power it takes to put roofs over our heads and meals on our tables. And we still rely on work to provide most people with most of the purchasing power they require to live.

It is the intersection of the flow of digital wonders and the reliance on work as a critical social institution that creates the possibility of a very difficult, very protracted period of economic discomfort. The next chapter will examine how existing social and economic institutions are managing the disruption of the digital revolution, and where the strains are most likely to lead to fractures.

2

Managing the Labour Glut

In a Volvo plant in Gothenburg, Sweden, acres of space are given over to the robots. In the final assembly area, there are typically teams of workers, in bunches of threes and fours, inserting smaller components into the nearly finished vehicle and checking to make sure previous steps were done correctly. As one walks back towards the beginning of the line, however, one is increasingly in the company of machines alone. Now and then a technician swishes by on a cycle, keeping an eye on the goings on. Automated production lines bring body pieces together to be welded in place on the chassis by robotic arms: four deft hydraulic limbs working in synchronicity in a highly choreographed set of motions. The combination of power and delicacy is extraordinary, yet the manufacturing process is remarkably flexible. I wondered aloud to my tour guides whether it was an annoyance to have to make vehicles for the British market, with the steering wheel on the right side of the car. Not in the least, was the reply. One vehicle on the line can be completely different from the next, from model type to detailed finishings. The equipment herds the right parts in the right order to the robots, which have no trouble at all going from a compact to an SUV to a sedan.

Robots in car plants are old news, though the latest machines are more sophisticated than ever: capable of building cars to an individual buyer's specifications and carrying out tasks while other robots simultaneously do their work in close quarters, in an elegant, slightly unnerving dance. What is newer is the work that is done a few miles away, in a rather ordinary office building elsewhere on the Volvo campus. There, highly skilled engineers write much of the code that runs the manufacturing process. They experiment with different production-floor layouts using a virtual model of the facility put together from a detailed laser scan of the actual production line. With a few keystrokes they can see whether rearranging the enormous machines will save time or leave robots banging their metal arms together.

Today, automobile manufacturing is first and foremost a software business, as opposed to an industrial operation. The value of the code in the machines becomes relatively more important as cars get smarter; Volvo, like many manufacturers, is working to get autonomous vehicles in regular operation on Swedish streets within the next few years. Already the cars are smart enough to do much of the brainwork involved in driving, from plotting routes to keeping a safe distance from the car ahead.

Driverless cars are not yet generating discomfort among the men who drive cabs around central Gothenburg, many of whom are immigrants or the children of immigrants. The hollowing out of the industrial workforce is, however. Income inequality has risen in this famously egalitarian country,[1] and recent Swedish governments have reformed their country's generous welfare programmes to encourage more unemployed people to seek work. In a country in which people with an immigrant background (that is, who are either immigrants or the children of immigrants) represent a disproportionate share of claimants, political support for generous welfare has broken down. Nor are the unemployed workers themselves especially happy about their lot; in 2013 a wave of unrest broke out as jobless youths – including both immigrants and Swedes of a far-right bent – took to the streets to vandalize property and pick fights. Yet the alternative – to push more people into the labour force to look for work – could exacerbate the problem of weak wage growth, especially for workers with lower skill levels.[2]

Who benefits from technological and economic change? The riches of high-income market economies are built on a foundation of creative destruction. Over and over, across the last two centuries, people and firms have developed clever new ways of doing things that displaced older ways, and made expendable the workers who practised them. For the benefits of economic change to be reasonably broadly felt, the workers displaced by robots or software must find a new economic niche.

The hope is that other industries or occupations will expand to absorb the displaced labour. But the process through which workers are reallocated from declining industries to growing ones has never been a pretty one. There is no central authority gently guiding people from fading forms of employment to more promising ones. There is no iron law that says that new, more profitable firms will create exactly enough of the right kinds of work to absorb those kicked out of shrinking occupations. On the contrary, displaced workers are quite often in an unusually bad position to be re-hired. They have spent years, or decades, accumulating know-how of declining value: such as how to use obsolete equipment or how to operate successfully within the culture of defunct firms. They often live in the wrong places too. For a laid-off manufacturing worker in Gothenburg, job openings at software firms in London are of little comfort.

Highly skilled workers, such as the engineers who use software to design new automobiles and plant layouts, have been made vastly more productive by new technologies. That productivity is built on a level of education and training that can't easily be attained by displaced factory workers, or indeed by most workers. And so the many, many people of modest education or training who have been displaced by machines are forced into competition for low-skill work which can't – for now – be done by machines. The glut of people angling for such jobs holds down wages and widens inequality.

It is tempting to believe that this balance of demand and supply for various types of workers is somehow unnatural, that were the economic decisions taken by governments more fair and less tilted in favour of the rich and connected, then labour markets might look more like they did in the past, when employers hoovered up cities full of less-skilled workers to do jobs that paid respectable wages. But that is a pipe dream. Policy has in many ways shifted in favour of the 'haves' rather than the 'have-nots', adding fortune atop good fortune. But the less comfortable position in which workers now find themselves is mostly due to structural change in the economy. The proof is in the paycheques: which, for a remarkably large share of the working world, have scarcely grown over the last fifteen years.[3]

Historically, growing economies dealt with labour by mobilizing less-skilled workers into high-productivity jobs, and by educating many others to equip them to meet the growing demand for highly skilled labour, leaving only a small share of workers competing for unproductive work at very low wages. But technological shifts mean there is much more labour around today, and fewer ways to mobilize that labour into high-productivity work of any sort, skilled or unskilled. As a result, a massive and growing share of the workforce is left to linger in the third category, accepting low pay in order to find employment in low-productivity jobs.

THE BYGONE AGE OF MASS EMPLOYMENT

Things were different early in the industrial revolution. Many of the key technologies of the nineteenth and early twentieth century were built on the mass use of relatively unskilled labour to produce valuable goods. Factory work, as awful as it often was, was nonetheless a powerful draw to people living in the countryside, struggling to keep themselves fed working on the land.

Early industrial advance often relied on the displacement of people by machines. Craft workers earning good wages, such as skilled weavers, found themselves put out of work by fancy new equipment that could produce much more cloth much more quickly and cheaply. These machines were extraordinarily productive; they made it possible for England to produce textiles in vast quantities, and to profit handsomely even as prices for clothes and cloth fell. But they could not operate themselves, these machines. They required human controllers. Not massively clever ones. Just ones that could be instructed – or programmed, if you will – to manage the simple tasks needed to keep the machinery running as it was designed to. Human workers became a part of the industrial machine, providing the oversight and direction that might today be handled by software. The indispensability of the human control mechanisms meant that as demand for the machines rose, so to did the demand for the human labourers needed to keep them humming.

The symbiosis between unskilled worker and machine reached its apotheosis in the assembly-line-driven plants of the early twentieth century. In the early days of the car industry, production was slow, expensive and laborious. Machine shops generally built the parts that the automakers needed, and the automakers employed skilled crafts-men, who often had to shape these individual components to fit the peculiarities of a car's handmade frame. In 1908 the Ford Motor Company sold only about 10,000 vehicles. Most of its 450 employees at the time were highly skilled mechanics and craftsmen. At the time, Ford bought most of the parts used in his cars from suppliers. The trained mechanics would then go to work on the parts, reshaping them to fit each automobile: cutting, smithing and welding, repeat-edly, in a process that was slow and very expensive.[4]

Henry Ford was famously determined to wring inefficiency out of this process. He settled on one design of automobile and mass-produced identical, interchangeable parts to a high degree of precision. He then borrowed an idea from the meat-processing industry. At the time, meat packers in Chicago worked along a 'disassembly line'. Carcasses hanging on hooks attached to a powered belt travelled past successive butchery stations. At each, lines of cleaver-wielding workers hacked off specific cuts of meat. As the animal moved through the factory its carcass grew smaller and smaller, while the meat removed from it was packaged and prepared for sale. Ford reckoned the system could easily work in reverse, with parts moving towards a powered line on which hung automobile bodies made ever larger by the crews that manned stations along its path. Ford invested in the machinery needed to move parts through the factory. Work crews were arranged in stations in an order optimized to boost efficiency. Powered lines then carried the chassis through production stages while other lines carried components to the stations where they would be added to the frame. The first modern assembly line.

With the arrival of the assembly line the labour-intensity of car production plummeted: from the more than 400 working hours needed to produce a car in 1909 to fewer than fifty hours two dec-ades later. Over the same period the price of a car, adjusted for inflation, fell by an estimated 80 per cent, while production of Model Ts rose from just over 10,000 per year to as many as two million by the mid 1920s. The fall in the price of a car was so great, and the consumer response so large, that employment in car production exploded in the 1920s even as the labour needed to produce each car tumbled.

The people working on the line were not especially skilled, for the most part. But Ford's clever system meant that they were nonetheless fantastically productive, enough so that Ford could afford to pay them well. He did; not for reasons of selflessness, but to reduce turnover in a plant in which the monotony could be overwhelming. As one worker eloquently described the feeling of working for Ford, 'If I keep putting on Nut No. 86 for about 86 more days, I will be Nut No. 86 in the Pontiac bughouse.'[5] Yet many workers kept on affixing nuts, despite the mental strain, after Ford introduced a $5-a-day wage in 1914, which more than doubled the previous pay rate, and which was accompanied by a reduction in daily working hours.

The use of relatively unskilled labour to make productive machines go was not limited to factory floors. As corporations of all sorts grew in size and complexity, their profits depended upon the flow of vast amounts of information: of payrolls and inventories, for instance, or customer accounts and revenues. Corporations therefore built up huge clerical operations to collect and process this information. Rooms full of secretarial and clerical workers typed and filed reports, managed the flow of memos around the business, and handled the calculations needed to track operations and keep the books. Much of this work was cognitive in nature – totting up sums, for instance – but highly routinized. The big macro process – running a global business – required lots of modestly skilled workers doing simple tasks.

There is precious little of that sort of work being created today. The digital revolution has its echoes of the old model, mind you. Uber, for instance, is a rough analogue. Traditional cab drivers are more like members of old craft guilds than you might initially think. Their jobs are protected by law and regulation (such as the medallions one needs to operate a New York City yellow cab) and special expertise that until recently had real value (like 'the knowledge' of London's tangled street grid one must obtain before operating a black cab). Uber entered markets with a new business structure that took advantage of technology – smartphones equipped with GPS – that made that prior knowledge (and 'the knowledge') much less important and valuable, and which made the process of getting a cab easier and faster for users.

In doing so it allowed relatively unskilled drivers to enter the business in vast numbers; many more people can operate a smartphone than can learn the entire maze that is London. It routinized and deskilled the labour involved. The cleverness of the technology at work and the business model are such that the cost of cab rides to users is often lower than the cost of taking a traditional cab, while Uber drivers, according to one analysis at least, earn more money per hour than traditional drivers: about $19 per hour compared to roughly $13 per hour for taxi drivers as a whole. (Cheaper cab rides can occur alongside higher wages because Uber's technology allows drivers to use their time more effectively.)[6]

The parallel is not perfect, however. Uber's success rests on the clever sidestepping of taxicab and employment regulation (tricks that have earned it significant legal scrutiny and which may not survive sustained legal challenges). Yet the firm's business does demonstrate how the technological deskilling of an occupation can lead to both a better experience for consumers and better pay for some workers.

Yet the example is not especially cheering. Many more of the digital revolution's disruptive business models work by reducing employment of less-skilled workers than by creating new opportunities for them. Other labour-intensive apps – such as TaskRabbit, which allows users to hire people for short-term gigs as errand-runners – work not because they make unskilled labour vastly more productive, but because unskilled labour is abundant and cheap enough to make it economical to harness such workers to do unproductive jobs: waiting in queues, for example.

Perhaps more importantly, new business models that open up opportunities for unskilled workers by simplifying the tasks done in an industry arguably pave the way for the eventual automation of those tasks. It would have been impossible to achieve present levels of automation in car factories in a world in which vehicles continued to be made by craftsmen smithing individual parts to a bespoke fit. Even the most advanced robots struggle to walk over uneven terrain; a cluttered workshop would be impossible for machines to navigate (and the rest of the job – the smithing and so on – would have been just as problematic). But once the process of building a car was broken down into many, very simple routines, automation became a snap. Robots can't walk around a messy room, but they can be programmed to make a precise series of welds over and over and over again.

Uber is helping to make the occupation of taxi driver automatable, by turning many parts of the job – spotting a would-be passenger, figuring out the route, handling payment – over to an app, making the driver nothing more than a vehicle operator. With carmakers and tech firms making significant strides regarding the automation of *that* role, there will soon be nothing left for cabbies to do. Uber's PR materials like to point out that the service is great for human drivers, offering them access to flexible, well-paid work. To investors, meanwhile, Uber emphasizes its desire to be a pioneer in the development of autonomous cab fleets.

Workers are expensive and troublesome. Firms that can routinize work make dealing with human labour easier by opening jobs to unskilled labour, of which there is an abundance, reducing worker bargaining power. They also take a significant step towards eventual automation: one that is becoming ever easier given the increasing power of digital technology.

The model of employment growth in which less-skilled workers operate highly productive machines has therefore played a small role in the creation of employment for new workers joining the global labour force, or for those displaced amid recent economic shifts. That small role will probably shrink in future rather than grow. Instead, two other employment-generating processes will dominate future labour markets.[7]

EDUCATION AS A RESPONSE TO CHANGE, AND ITS LIMITATIONS

The industrial revolution didn't eliminate skilled workers. Many craftsmen were wiped out by new manufacturing techniques, but industrial economies quickly developed a near-insatiable need for better-educated workers. The factories sweeping aside smiths and weavers required skilled scientists and technicians: chemists, metallurgists, and mechanical and electrical engineers. As the revolution wore on and the state of the industrial art advanced, the sophistication of the expertise needed on factory floors grew tremendously. Industrial chemical plants could not be left solely in the hands of workers with a secondary school certificate and a 400-page manual.

Office buildings also filled up with highly skilled workers. Management became more important. As operations grew more complex and spread across borders, the need for trained lawyers, accountants and financial officers exploded. Companies suddenly needed to manage the effects of things such as currency risk and international financial and accounting rules on the flow of profits across borders. Firms also began taking a more sophisticated approach to marketing and public relations: steps that further increased the demand for highly skilled workers.

The workforce of the early industrial era was not exactly ready to waltz into the laboratory or the executive suite. In the early nineteenth century, few people were equipped for office life. Most were accustomed to a rough rural existence that required more elbow grease than social grace. Most were illiterate and innumerate. Mobilizing such workers from the farm to the factory was one thing; getting them from the factory to the cubicle was quite another.

The key was education. Social reformers pushed campaigns to broaden public provision of primary education, helping to transform industrializing societies. For the first time, the great mass of people was taught to read, write and do simple arithmetic. In the late nineteenth and early twentieth centuries, governments pushed for universal secondary education. They also took steps to broaden access to higher education. By 1940 roughly a quarter of working-age

Americans had at least a secondary school education and around 5 per cent had at least a bachelor's degree; rates were higher for the younger cohorts.[8]

Those figures rose steadily over the next half-century; now nearly 90 per cent of working-age Americans have at least a secondary education and 41 per cent have a bachelor's degree or more. Most other rich countries do about as well; about 39 per cent of Britons have a bachelor's degree or better, as do 26 per cent of Germans and 46 per cent of Japanese (The average among OECD countries is about 30 per cent of the working-age population).[9] Humanity spent millennia figuring out ways to augment its physical strength, through wheels and pulleys and animal-power and steam and electricity, but, in the space of just over a century, humanity suddenly mobilized an enormous share of its *cognitive* strength.

Rising skill levels enabled rapid economic growth; the second industrial revolution, built on technologies such as electricity, chemistry and the car, couldn't have unfolded as it did without a growing pool of skilled labour. It wasn't just the workers in the labs that mattered either; America's rapid, resource-intensive growth also owed much to the strength of the emerging mineral industries built on geologic expertise, cultivated at specialized institutions (such as Columbia University's School of Mines). In the early twentieth century, America was the world's top producer of almost every industrial mineral that mattered.[10]

But rising skill levels were also helpful in improving the distribution of growth. In 1850, only about 5 per cent of employment in America could be categorized as highly skilled (meaning in professional, technical or managerial work). That figure rose to 12 per cent by 1920 and to a third of all employment by 1990. High-skill work could expand to account for so much employment because qualified workers were available to fill the positions; expanding education meant that each new labour-market cohort was better educated than the last. And because the workers were available to fill those positions, the share of Americans labouring in the best-compensated sorts of jobs rose – and, in the same way, the share competing for positions requiring less education and paying lower wages fell. More workers had degrees and were earning the higher wages to which college graduates had access.[11]

And because the supply of graduates was growing so rapidly, the wage boost that a worker received from completing a degree tumbled by about half from 1910 to 1950. Wages were rising rapidly and the wage distribution was narrowing. In other words, it was good for highly educated Americans that they were able to complete degrees in large numbers and go on to work as engineers and accountants; yet because so many Americans were going to university, the glut of less-skilled workers competing for jobs on factory floors, or painting houses, or cleaning offices was reduced. Firms had to work a little harder to fill those positions, and that meant faster growth in the wages of the people that held those jobs.

Unfortunately, the skill-upgrading approach to more and better employment, which worked so well for most of the industrial era, has run out of steam. Cracks in the facade were apparent by the early 1980s in America, when growth in the share of people obtaining a university degree levelled off; in Europe, which long lagged behind America in educational attainment, this plateau occurred later. But evidence from across the rich world suggests that it is simply very difficult to boost secondary-school completion rates above 90 per cent and to raise university completion rates as high as 50 per cent.[12]

University is hard. Many of those who don't currently make it through a college programme lack the cognitive ability to do so. Others could be helped through with better preparation and more attention. But the human and financial resources needed to improve the performance of less-prepared students mean that it is much harder and more costly to raise college completion rates from 40 per cent to 45 per cent than it was to lift them from 20 per cent to 25 per cent. In the absence of magical new innovations in education or cognitive science, the populations of advanced economies are close to being as educated as they can reasonably be expected to be. That means that the proportion of highly educated workers to less-educated workers is no longer going to rise in the growth-boosting, inequality-dampening way it once did.

At the same time, the demand for skills has continued to evolve. Through the last two decades of the twentieth century, the appetite for university graduates grew tremendously even as the supply of new graduates stalled. As a result, the wage premium earned by graduates rose back to where it had been early in the twentieth century. In America, around 1980, the typical college graduate earned a wage about 40 per cent higher than that of the typical high-school graduate. By 2000, that premium had nearly doubled. The degree premium is highest for those with technical or scientific training. Where the typical graduate with a degree in English literature might make 50 per cent more per hour than a high-school graduate, an economics graduate might earn double the high-school wage, and an electrical engineer 2.5 times the high-school wage.[13]

Since 2000, the demand for skilled workers has shifted towards those with advanced degrees. College graduates still earn a healthy premium for their degree, but not because wages for those with bachelor's degrees alone have been soaring. Rather, pay began to stagnate for college graduates much as it had for workers with lower skill levels in the late twentieth century. Completing college remains a ticket to higher pay, generally speaking, but not to rapidly growing pay, as in the past.

Rapid growth in incomes keeps receding to higher and higher echelons of the income and skill distribution pyramid. The typical worker with an advanced degree earns about 30 per cent more than the typical worker with a bachelor's degree alone, and that gap continues to rise. An advanced degree is most lucrative for those in fields such as engineering and computing, finance and economics; the modern economy delivers its biggest salary rewards to those who build the technologies and finance and manage the companies of the future. Only about 10 per cent of American adults have obtained a postgraduate degree of any sort. That share will almost certainly rise over time, but it is unreasonable to expect most university graduates to go on to complete extremely difficult advanced degrees in engineering or economics.[14]

Just as most advanced economies are reaching the point at which it becomes difficult to improve educational attainment any further, the level of education needed to participate in the most lucrative corners of the economy is growing beyond the reach of the vast majority of workers. Indeed, one of the troubling dynamics within labour markets over the last fifteen years has been downward mobility of college-educated workers; those with degrees have often been forced to take work for which they are overqualified, in the process pushing those with less education into competition for even less skill-intensive, and less lucrative, work.

Skilled workers today are exposed to the same disruptive forces that battered less-skilled workers over the last generation: automation, globalization and rising productivity. The emerging world has its billions of brains. Emerging markets are producing growing numbers of engineers and doctors. These workers migrate to rich economies when they can; they compete in rich economies using technology when they cannot. Hospitals send scans abroad for examination and diagnosis. Coding is also 'offshored'. More of this will be possible in future.

At the same time software is becoming better at doing some skilled tasks, such as writing reports or analysing documents. And other technological developments are allowing top performers in finance or media or education to serve a huge clientele. The low-hanging educational fruit has been picked. And growth in opportunities in moderately skilled positions is struggling to keep up with the effective labour that can be applied in those industries. Therefore, less of the burden of adjustment to technology can be borne by education this time around. Consequently, much more of the burden will fall on a third mechanism: falling wages.

FALLING WAGES IN THE AGE OF ABUNDANCE

As the world sank into financial crisis and recession in 2008, consumers lost interest in spending on all sorts of things, from cars to meals out. Tumbling demand forced firms to tighten their belts; faced with awful business conditions they could not keep producing as usual and expect to stay in business. In America, companies responded by sacking lots of workers. Employment fell by nearly nine million jobs, or more than 6 per cent, between 2008 and 2010. In Britain, by contrast, firms also slashed their payrolls, but by much less than in America: about 2 per cent.

Britain didn't send fewer workers packing because its recession was any milder. On the contrary, British GDP fell by much more than America's did: by about 7 per cent compared to a 4 per cent decline in the United States. So why did British companies find it so much easier to hang on to their workers?[15]

The answer comes down to what happened to pay. In Britain, real wages fell off a cliff, declining by about 8 per cent over the course of the recession and early recovery. In America, on the other hand, wage growth slowed sharply but remained positive, on average, between 2007 and 2013. In Britain, because workers got much cheaper as sales plummeted, bosses could afford to keep workers on and work them less hard; productivity in Britain also dropped sharply alongside sales and wages. But in America, workers weren't getting any cheaper even as business was declining dramatically. Bosses therefore had little choice but to fire lots of people in order to stay afloat, and to work the ones they kept as hard as they could; productivity in America rose sharply during the recession and early recovery.

When workers are cheap, companies have much more flexibility in choosing which people to hire and retain and how they use them. When workers feel a pinch, because a recession has led to tumbling economic activity or because new technologies are adding massively to the amount of labour competing for work, falling labour costs become one of the most important mechanisms through which most willing workers find or stay in employment. From the workers' perspective, falling wages are hardly ideal. Low or falling pay is dis-piriting. It forces households to make difficult choices and leads to reductions in living standards. It can, in fact, lead to slower long-run economic growth. Yet when labour is extraordinarily abundant, and when workers have no choice but to seek jobs to provide for them-selves and their families, the downward pressure on pay can become intense.

Cheap labour can facilitate employment growth in a few different ways. Low wages can encourage people to use more of some kinds of manual or service labour. As pay for low-skill workers stagnates, for example, more households might find it attractive to hire a house-cleaning service or a landscaping firm, to get nails done at a salon rather than at home, or to retain the services of a personal trainer. The more labour is available at very low pay, the more extensive this low-pay service economy can become. In fact, we have a very good idea what mass employment in low-skill service work looks like, thanks to examples in poorer economies, where crowds of attendants work at dubious productivity levels in hotels and restaurants, and in the eighteenth and nineteenth century, when domestic payrolls, Downton Abbey-style, absorbed large numbers of workers.

Low wages can also boost employment by discouraging firms from automating. Industrial manufacturing in parts of China and India uses many more workers than similar processes would in Europe or Japan, where labour costs are much higher. When wages are low enough it doesn't make sense to replace cashiers with an automated checkout, or to use robots in logistical tasks in warehouses. Indeed, if wages fall by enough, firms may actually replace some automated processes with human labour. There is some evidence that this occurred in Britain during the recession, especially in the service sector: that firms relied on workers to do jobs that might otherwise have been managed by maintaining and upgrading computer and software equipment. Law firms delayed investment in digital document-management systems because skilled legal assistants could be retained for a song; contracts with expensive data-crunching companies were allowed to lapse, because teams of cheap analysts could be brought on to do the work in-house instead.

When things are abundant, they are used carelessly. When water is plentiful, people leave taps running and irrigate massive, thirsty lawns hours after a rainstorm. When labour is plentiful, three workers pour the tea. If labour abundance is dramatic enough and prolonged enough, then the entire structure of on economy can warp, as firms put people to work doing low-value kinds of tasks. Investment incentives change. Growth patterns change. And the gap in incomes and in satisfaction between those doing necessary work and those who reduced their wage demands until they found a job – greeting shoppers, say – increases.

Evidence suggests that this third mechanism is playing an increasingly important role in the process of squeezing workers displaced by the digital revolution into new jobs. Workers in most industrializing countries experienced enormous rises in income between 1870 and 1970. In America, for example, between 1947 and 1972 the average real wage grew by between 2.5 per cent and 3 per cent per year, at which pace pay, adjusted for inflation, doubles about once every three decades. Since the 1970s, however, increases in real pay have been disappointing. In America real wages have since grown by less than 1 per cent on average each year, a rate at which it takes just over seventy years for pay to double. Even during the extraordinary boom from 1994 to 2005, wage growth, adjusted for inflation, was no more than around 2 per cent per year.[16]

Productivity growth followed a similar trajectory – growing rapidly until the early 1970s and then performing poorly but for the spurt from 1994 to 2005, yet it nonetheless performed better than wages. Prior to the early 1970s, productivity growth was rapid *and* real pay growth largely kept pace with productivity improvements. That is, as workers got better at generating output, the fruits of those improvements accrued to the workers themselves, in the form of higher pay. Thereafter productivity growth slowed *and* workers failed even to capture the benefits of that slower growth; from 2005 to 2014, for instance, productivity grew at about 1.4 per cent per year, or about twice as fast as growth in real wages.

As bleak as these numbers are, the focus on averages presents too rosy a picture. Median wage growth, or growth in wages for the American worker in the middle of the distribution, did far worse. Indeed, since 2000 the real wage for the typical American has not risen at all. Looking further back does not much improve the picture either; since 1980 the median real wage is up by only about 4 per cent. Not per year, but over the whole of the period. And if you then focus in just on the real wage of the median *male* worker, the duration of the stagnation extends back into the 1960s.[17]

America is not an outlier; on the contrary, its performance looks better than that of some other rich economies. The real wage of the typical Briton, for example, did much better up until about 2008, but has since fallen dramatically, to an extent unmatched in any other large rich economy, while, from 1995 to 2012, the average real wage in Germany, Italy and Japan all underperformed that in America; indeed, in Japan the real pay actually fell.[18]

This dismal performance can't easily be explained away. Of course it is true, as some note, that wages and salaries have come to account for a smaller share of total compensation over the last half-century, as more of the compensation package has been accounted for by benefits, such as pension supplements and healthcare coverage. In America those benefits accounted for nearly a quarter of total compensation in 2013, up from about 7 per cent in 1950. Yet this does not really change the picture. On the one hand, much of the growth in benefit compensation has come from rising health insurance contributions, and since that is driven by soaring healthcare costs, it hardly represents much of an improvement in inflation-adjusted compensation. More importantly, growth in total benefits has also stagnated for much of the last twenty years.[19]

The hardships suffered by workers show up in other worrying trends as well. One is rising income inequality – which helps to explain why average incomes have risen faster than median ones, since those at the top have risen most. Though inequality is occasionally dismissed as an American problem, dispersion in incomes is, in fact, widespread. Over the last thirty years, the share of total income earned by the top 10 per cent of earners in America has soared from about a third in 1980 to half today. Many other economies – Britain, Germany, Italy, Japan and Sweden among them – have also seen more income flow to top earners, even if the rise has not been as dramatic as in America. Perhaps most striking, inequality has also been climbing in fast-growing emerging markets such as China and India. Inequality globally has fallen, however, in recent decades, as poor countries have grown faster than rich ones, but within *both* rich and poor economies inequality has mostly risen, and the rise shows few signs of abating.[20]

A similarly disconcerting, and widespread, phenomenon is the decline in the share of income which flows to workers, as opposed to owners of other factors, such as capital and land. This 'labour share' was, for much of the twentieth century, assumed by economists to be roughly constant over time. A stable labour share was one of six 'stylized facts of growth' set out by the renowned University of Cambridge economist Nicholas Kaldor in 1957.[21] Yet since about 1990 labour share has trended down globally. Some research suggests that the decline would in fact have been larger than it was were it not for the rapid increase in top incomes. This decline, incidentally, is what we would expect in a world in which productivity is growing faster than wages; the difference is captured by someone, and if it isn't workers then it is some other group with a claim on an economy's economic output.[22]

It is not a coincidence that these trends all developed at roughly the same time, in the 1970s and 1980s. They represent a distinct break from what had come before. For decades before that, real wage growth kept up with productivity growth, which had itself risen faster than in any prior period. Income inequality, which had been extraordinarily high in the early twentieth century, fell dramatically from the 1930s to the 1950s and stayed low for the two decades after that. And, before this period, the labour share 'wiggled' yet did not trend, not as it has over the last generation.

So where does this leave us? When workers are displaced from one set of tasks, some go on to compete with other highly skilled workers to do cognitively complex tasks – but most don't. The truth is that while there are plenty of challenging tasks at which humans have a significant advantage, from writing poetry to building new economic theories, most human workers are also unable to do such work effectively. Huge investments in education may improve the employment outlook for some workers, but no amount of education will allow the typical worker to contribute at the frontier of scientific discovery.

Instead, most displaced workers fall into competition for tasks requiring low skill levels. As the supply of workers seeking employment in such tasks grows, wages fall. That, in turn, encourages firms to use more human labour – and, paradoxically, to take less advantage of the possibilities of automation than they could. In other words, technological progress and productivity growth have been self-limiting; rapid change in some parts of the economy displaces millions of workers, leading to lower wages, more employment, and economic stagnation in less skilled parts of the economy, which will expand like a sponge as they absorb ever more cheap labour.

Should this continue, the implications for social stability will be significant and worrying. But will it? The next chapter considers whether technological change itself might mitigate the downsides of labour abundance.

3

In Search of a Better Sponge

Jobs appear unexpectedly. There were times, in days gone by, when it would have been impossible to anticipate the looming need for lamp-lighters, telegraphers and social media strategy coordinators. Humility in the face of technological change, regarding what employment opportunities that change might deliver, is generally a sensible attitude.

Yet while it might be hopeless to try to say, with any certainty, what new occupations will appear in the near future and amid the march of digital technology, we can attempt to understand what new forms of work would need to look like in order to qualify as opportunities for mass employment – and take a guess at how likely they are to fit the bill. Technology may surprise us – it often does – but the outlook for mass employment in productive, well-compensated jobs looks dim.

The problem is the sheer abundance of labour. Work in highly productive jobs cannot grow to absorb a large share of available labour without creating a glut of the product or service being produced, driving down prices and constraining further growth (both in the industry itself and in the wages paid to workers).

Meanwhile, the technological capacity to eliminate costly labour will continue to improve. That capacity will tend to be directed towards industries where labour accounts for a large share of production costs. Highly productive, well-compensated forms of mass employment either sow the seeds of their own elimination or stick out like great, expensive sore thumbs, begging to be swept aside by technology. New technologies will create new, good work, which might often benefit the less skilled. But it will not be scalable mass employment. And it will not solve the problem of labour abundance.

ROBOTS WITH BLUE COLLARS

Astronauts residing in the International Space Station can watch the world's great cities slide past them as they glide over the continents on the night side of the planet. Beneath them, as they cross North America, the lights of cities sketch out the familiar geography of the metropolitan United States. But, in recent years, ISS visitors have noted an oddity in the picture below. At the Mississippi River, the great tangle of lights of the eastern cities gives way to the dark of the prairie. But there on the northern Plains, west of Minneapolis and north of Denver, where nothing but emptiness ought to be, is a blaze of light as big as Chicago.

What has taken over the North Dakota countryside is not a massive new supercity but the fracking wells of the Bakken shale, one manifestation of an extraordinary American energy revolution. The hundreds of wells that dot the land are spot-lit at night, and are occasionally ablaze with light when excess natural gas from the wells is burnt off.

Of the new work that resembles the mass employment of the industrial past, jobs in fracking are probably the closest analogue to industrial-era factory jobs. Hydraulic fracturing (fracking) has, in fact, been around as a technique since the middle of the twentieth century. But innovations to the process, including a move to horizontal drilling, opened vast shale deposits to development at a time when the global oil price was rising dramatically. The result was an extraordinary boom in oil and gas production, centred on American shale deposits.

American production of oil and petroleum liquids, which entered a steep and steady decline in the 1980s, has more than doubled since 2008, to about fourteen million barrels per day in 2014, making America the world's largest producer of oil, ahead of Saudi Arabia.[1] The boom generated a jobs bonanza. From 2010 to 2015, employment in North Dakota, one of the focal points of the shale revolution, rose by about 30 per cent (as did nationwide employment in oil and gas extraction), compared to an increase of about 8 per cent for all US employment.[2] In a 2012 speech, Barack Obama estimated that fracking could employ as many as 600,000 people by 2020, most of them blue-collar workers.

As many as 600,000. That is less than half the number of Americans now employed in trucking, many of them blue-collar workers, whose jobs may soon be put at risk by automation – in October of 2015, Mercedes Benz road-tested a fully autonomous truck on the Autobahn near Stuttgart.[3] And yet even the current hopes for blue-collar employment in fracking now look wildly optimistic. Work in oil and gas extraction in America rose to just over 200,000 in late 2014, but the resultant rapid growth in oil supply drove prices down, starting in the summer of 2014, from the roughly $100 per barrel level that prevailed between 2010 and 2014 to around $50 per barrel. The glut has led to a sharp decline in the drilling of new wells and therefore in employment. Owners of the wells that continue to operate have begun looking for ways to cut labour expenses. When fracking investment rebounds, as it inevitably will, the industry will have, by then, found ways to make itself less labour-intensive.

The employment opportunities of the future will be profoundly constrained by the capacity to automate work, and by the abundance of labour. Those two forces will combine to generate an employment trilemma: new forms of work are likely to satisfy at most two of the following three conditions: 1) high productivity and wages, 2) resistance to automation, and 3) the potential to employ massive amounts of labour. Fracking jobs pay good wages, because the value of the goods being produced is high, and the work, for now, can't easily be automated away. Unfortunately, the work cannot scale; the moment employment grows exponentially, the resultant soaring output then depresses oil and gas prices, curtailing further growth.

The same trilemma faced by the fracking industry will almost certainly constrain other forms of work commonly offered as potential sources of blue-collar employment in the future. Consider 'green jobs', for example. Within that category there are occupations that are both high-productivity and scalable, such as work on production lines making wind turbines or solar panels. But, unfortunately, such work is easily automatable. Since 2000, for instance, the cost of solar panels has plummeted as production has soared. Solar-panel manufacture in China has grown immensely, rising from 50 megawatts of capacity produced in 2004 to 23,000 megawatts in 2012. That extraordinary growth contributed to tumbling prices; the cost of solar panels fell by half from 2011 to 2014. Falling prices then proceeded to squeeze Chinese producers, who have turned to automation to hold down production costs. Many new solar-panel production lines are now fully automated, save for a few humans conducting quality-control checks.[4]

Cheap solar panels have, of course, increased interest in installation of home solar-energy systems. Installation of these systems means work for people with modest skill levels. But the extraordinary decline in the cost of solar panels means that most of the cost of a solar-energy system is in the labour. That, in turn, puts a limit on how high wages in installation can rise. Should they grow too high, households will instead opt to draw from the grid; power companies can also avail themselves of the use of solar energy on a large scale, but building and maintaining a central solar plant typically requires less labour, with higher skill levels, than rooftop installation. America's Bureau of Labor Statistics (BLS) estimates that there are about 5,000 installers working in America now, at a good wage: the median income for solar installers, at $37,000, is above the median pay for all Americans. The BLS estimates that employment in the field could grow to 6,000 by the early 2020s, but that is a mere drop in the employment bucket. The numbers could, of course, rise, but they will be constrained by installer pay. The less such workers cost, and the more financially attractive rooftop solar looks relative to less labour-intensive alternatives, the more installers there will be – a difficult but increasingly common occupational trade-off.[5]

Installation work is resistant to automation. The future of the work is either one in which employment grows while pay stagnates, or in which the work becomes more productive – because much more solar energy is generated at solar-energy plants where the energy output per person is much higher – and employment stagnates.

There could be other, similar opportunities in different fields. Michael Mandel, chief economic strategist at the Progressive Policy Institute, invites those sceptical of the job-creating power of new technology to imagine a world in which doctors can 3D-print new organs. In that future, humans would spend lots of time swapping out worn out livers, say, for new ones, and would need basic nursing care at every operation. Mandel might be right, but that vision of future work relies on a very specific version of biomedical advance; innovations that grow the organ on the inside of the body or repair existing organs non-surgically might, alternatively, dramatically *reduce* the need for medical care.[6]

Indeed, while fields such as education and healthcare have long been held out as the great hope for future employment growth, that hope is built on an assumption that productivity in those industries will remain low. But it might not; the future of work in education and healthcare hinges on how society opts to resolve the trilemma.

COST DISEASE, AND THE DOWNSIDE TO JOB CREATION

William Baumol is an American economist. His career has been a long and productive one: he finished his PhD in 1949 and published his most recent book in 2012. Yet among his most significant contributions to the world is the story behind stagnant productivity growth across large swathes of modern economies.

Many service industries, including critical sectors such as education, are subject to a phenomenon known as Baumol's Cost Disease. Pay, economists reckon, ought to correspond roughly to productivity: the more productive a worker is, the more a firm can afford to pay him. Yet this is not always how wages work.

As an economy grows and develops, some industries become much more productive. Workers in car industries, for instance, learn how to make better quality cars at lower cost, all while boosting the number of cars that can be manufactured. Higher productivity in the car industry translates into rising wages (remember Henry Ford?). But while the car industry or the electronics industry in the growing economy are getting better at doing more with less, other sectors are not. Waiters in restaurants don't go from serving six tables in an hour to serving 600. Barbers don't find ways to simultaneously give eight haircuts. Concert violinists can't play their concertos any harder than they already were, and dentists still find themselves hunched over one mouth at a time.

But in all these service industries, in which productivity is growing very slowly or not at all, wages also tend to rise over time; playing in an orchestra, for example, generally won't make a person rich, but violinists do manage to earn a bit more than their peers did in the seventeenth century. Wages for these jobs go up because the workers in them operate in the same labour market as the workers who trudge off to the car plant every morning. As wages rise in productive indus-tries, restaurant workers and hair cutters and the like quit their service-sector jobs and seek work at the factories. But this creates a scarcity of these kinds of workers – the restaurant workers and hair cutters – that can't be sustained; factory workers earning decent wages want to spend their money after their shift ends, on haircuts or meals out. Prices and wages for low-productivity work therefore rise until enough workers can be tempted back from the factories to sat-isfy the demand for haircuts and so on. Restaurants and salons are forced to raise salaries, though their employees haven't necessarily become more productive. Productive firms drag up the cost of living right across an economy.

Baumol's Cost Disease means that the cost of many critical sectors in an economy tends to rise over time, as the economy as a whole becomes richer and more productive. Hospitals have to offer doctors and nurses higher and higher salaries, even though those doctors and nurses aren't tending many more patients than they were a generation or two ago. Teachers still teach about as many students as they did in the late nineteenth century, i.e. a classroom full, but teacher pay – if lower than many people reckon it ought to be – is much higher than it was a century ago. Public sector employment in general tends to follow this rule: salaries must rise to remain competitive with those in the private sector, despite the fact that productivity in the public sector rises very slowly, if at all.

This dynamic is generally considered to be a bad thing. It is the reason that education and healthcare cost so much. Yet low produc-tivity, and the expense of these services, has a corollary: lots of jobs. Since 1990, total employment in America has risen by just over 30 per cent. Employment in both the education and the healthcare sectors has doubled, by contrast.[7] One vision of the future of work is that these sectors – education, healthcare and government – will continue to grow: soaring productivity in other parts of the economy will release labour that will be soaked up by the low-productivity sponges. But that is a dismal prediction of the future in many ways; it implies, for instance, that important public services never become much cheaper and more widely available. It might also be an unsustainable future, as growth in the share of national budgets spent on healthcare or bureaucracy tends to create pressure to cut costs and ration access.

But the digital revolution carries with it the potential to alter this dynamic. Consider higher education. The technology of the university has not changed very much over the last millennium or so. Now, as in the distant past, students gather together in a room to hear a scholar speak aloud lessons on mathematics or history. If a school wishes to accommodate more students, it can increase the size of the lecture halls, but before long it must add more buildings and more professors. Productivity growth within higher education has historically been almost nil.

Poor productivity growth in higher education has consequences. Employment in higher education, as mentioned above, has doubled in a generation. Its cost has risen steadily and dramatically, at greater than the rate of inflation. Rising costs have stressed both students and governments. Tuition fees have risen in America and have been introduced in other countries, such as Britain, where they didn't exist before. The cost of government subsidies has also risen. So has student borrowing.[8]

But despite all of this, universities are not obviously doing a better job of educating students. The share of students attending and finishing university has plateaued across the rich world; in America it has barely gone up at all since the 1970s. And, since 2000, as we have discussed, pay for college-educated workers in advanced economies has stagnated. The economic role of higher education in rich economies – in terms of employment and spending share – is going up and up and up, but neither the educational attainment of the population nor the returns on a degree are increasing.[9]

Rising costs for mediocre results have focused the interest of entrepreneurs and technologists on the problem. As online communication among and between students, tutors and professors has improved, and as the sophistication of online course materials has grown, models of online education have begun to emerge. These models are often referred to generally as MOOCs, which stands for 'massive open online courses'. In practice, there are many different kinds of MOOC, and many different educational forms that could grow up around them. But the MOOC is a very important development for the world of education.

A MOOC, generally speaking, is an online course that consists of online instruction and assignments, often interactive. A student can enrol in a course, work through video lectures or instructions, email questions to tutors or work through them within online discussion forums, submit completed assignments (often graded by other students), and then complete online examinations. Once a MOOC has been created, it can be offered, more or less, to as many students as are interested in taking it, and the cost to the institution offering the MOOC of the hundredth enrolee is not much different than the cost of the millionth enrolee: in both cases it's basically nothing. A student can take a MOOC anywhere in the world that there is internet access. A student can also work through the lectures whenever it is convenient: at night after work, over lunch, on weekends, and so forth.

It isn't hard to see how this might be transformative. Because the cost of an additional student is almost nothing, MOOCs work on a different economic model to traditional colleges and universities: the incentive for producers is to invest lots of money in the fixed cost of creating a high-quality MOOC, in the hope of attracting vast numbers of enrolees over which the costs of creating the courses can be spread. (That cost sometimes comes in the form of a fee to enrol, but is more often levied on students who wish to obtain a certificate of completion after successfully working through the course.) For students, that means it is very cheap and easy to try out courses, to experiment with different subjects or different offerings from different providers. It is cheap and easy to take a course multiple times, or to work through the courses leading to a degree a bit at a time, so it is therefore cheap and easy to supplement one's traditional education or one's work training with time spent on MOOCs.

No one single model of online education will meet every need or displace existing forms of higher education wholesale, though. MOOCs are better substitutes for some kinds of instruction than for others. But not all of higher education would need to go online for there to be massive disruption to the industry. The image most people conjure up when asked to picture a university might be an idyllic scene of Victorian buildings, with interested students engaged in high-minded discussion with attentive professors. In practice, the median university experience is something very different: a massive lecture course taught by a nondescript professor using bog-standard course material at a university that is not especially competitive (which does not, in other words, reject most applicants). And *that* sort of experience could very easily be swapped out for an online course with little loss in quality. Indeed, universities themselves are opting to make this swap in many cases by moving towards 'flipped classrooms', where students receive most instruction online, then come to a physical classroom for discussion and to work through difficulties with a professor or graduate assistant.

Over time, and with further experimentation, the quality of online courses will improve. In many educational contexts it will make sense to replace in-person lectures with these courses. The consequence of this replacement will be an educational experience that is probably just as good for most students as what came before, but which is substantially cheaper for students and universities, and which employs many, many fewer mediocre professors.

In a MOOC world, a handful of very good introductory economics courses, created by teams of top instructors and skilled producers, could make hundreds or thousands of intro-level economics instructors redundant. Those few teams will probably earn a lot of money, even as the total amount spent on instruction falls. Meanwhile, the academics who once earned a good living as lecturing professors may instead find themselves labouring on more tenuous and less lucrative contracts, tutoring students who need some in-person guidance as a supplement to their online work, while many of the less-skilled teaching assistants who previously did that work will find themselves pushed out of the industry. And many of the administrators and service workers that previously kept universities ticking will lose their jobs. To the extent that technology can be used to cure the Cost Disease, it will potentially yield a better, cheaper experience for consumers but many, many fewer jobs.

Healthcare, though not a perfect parallel, is likely to be affected by similar forces. More diagnostic work will be done at a distance (or automated), generating fortunes for those who perfect models of distance medicine. That, in turn, will force many doctors to adjust, to maybe take on very different sorts of work at less pay, setting in motion a process of downward displacement across employment categories in medicine. The in-person services people desire in healthcare will not especially be those services associated with expensive expertise; they will be those associated with bedside manner or emotional engagement, or with a willingness to do basic, manual and often unpleasant tasks. A world in which remote monitoring equipment in our smartphones could account for most of the interaction we now have within physical doctor's offices (but in which in-person health counselling or therapy is available when necessary) is, in many ways, a better one for patients. But it will be achieved by wringing massive amounts of inefficient labour out of the system.

Education and healthcare have been the great labour sponges of the last generation. Much of the promise of the digital revolution, however, lies in its potential to make these sectors better and more affordable. What that will mean is that a few will do the work that has until now been done, at great cost, by many. The employment trilemma implies that we can have high-employment public sectors in future, but only at the cost of low productivities and very big hospital, tuition and tax bills.

NEW ECONOMY, LOW-WAGE FOUNDATIONS

Does the world clearly need sources of *mass* employment for all those who would like to work to have a good job? One might think that the digital revolution offers the possibility of a better sort of work than was available to people with modest skill levels a generation or two ago. Perhaps the trilemma can be dispensed with altogether?

The web does indeed create interesting new niches in the economy by simultaneously expanding the size of the market and making it easier for people to find precisely what they are looking for within that market. A larger market increases the scope within the economy for specialization; when there are more potential customers available, producers do not need to appeal to as large a share of the total market to make a buck – provided that there are good ways for people to find the speciality product available for sale.

Consider this example. A few years ago two economists at the Massachusetts Institute of Technology, Glenn Ellison and Sara Fisher Ellison, went hunting for a book. It was an obscure book – a thirty-year-old text analysing the pharmaceutical market – which (shockingly!) was out of print, and which the MIT library did not have on its shelves. The team did what any modern economist would and took to the web, where a search at an online used-book site turned up a copy, on sale for about $20.

When the book arrived, they discovered, pencilled on and then incompletely erased from the inside cover of the book, a different price: $0.75. The old book had apparently lingered unwanted on some dusty shelf in a used bookshop, waiting in vain for a member of its niche audience to happen upon it. The bookseller had priced it rather optimistically at just a shade above zero in the hope that it might be worth at least that much to a passing customer. Online, however, the bookseller found someone who did want it, very badly. And because they wanted it badly, they were willing to pay much more than $0.75, and, indeed, much more than $20. The web enabled a transaction that benefitted both the used-book seller and the buyer – enormously so.[10]

Might it be possible to build a labour market in which workers specialize to a high degree, and then rely on the mass market created by the web to find a customer, or employer, willing to pay good money for that particular, specialized skill? Could human workers be like that pharmaceutical text?

In some cases, the answer is clearly yes. The specialization effect is easy to spot on YouTube, where there are video-gamers who can reportedly make six or seven figures through their personal channels, producing videos which walk players through new games and which generate phenomenal amounts of traffic. These video stars might have found their way into traditional media in a web-free world, or they might instead have lingered in the memories of their university roommates as the dudes who were amazing at video games and who cracked everyone up while figuring out how to beat the newest release.

Something similar is happening in the market for crafts of all sorts produced by hobbyists. Online marketplaces for craft producers abound. Etsy, for example, employs fewer than 1,000 people directly, most of whom are located in its Brooklyn, New York, headquarters. The site has more than one million affiliate sellers, however: independent makers of artwork, clothing, jewellery, craft goods, assorted trinkets and other curios, whose combined sales reached $1.35 billion in 2013. Etsy makes it possible for the person who made a hobby of creating sewing-sampler wall hangings with rock lyrics to find people who want to pay money for just such a product, and perhaps to sell enough as a result to earn a modest income.[11]

Some analyses suggest that these sorts of niche work could become part of a 'gig economy' that provides supplemental income and work for lots of people. That is, as 'regular' work, in well-defined jobs for large employers, provides workers with less wage growth and fewer hours, they will increasingly turn to a few hours driving an Uber, or a side business selling craft goods, to top up their income. In time, perhaps, the gig economy could become the regular economy; the flow of earning opportunities could grow large enough that workers could feel secure in their ability to earn a living through piecework. In emerging economies, the gig economy could permit workers to make the leap directly from the poverty of the developing world to full participation in global markets; a handful of residents of Mumbai slums have boosted their incomes tremendously through participation in a programme offered by eBay, which allows them to sell their wares (such as handmade leather goods) to customers around the world rather than to those in nearby Mumbai neighbourhoods.

How powerful could this gig economy become? It is growing every day, though from a very small base. Uber, one of the larger contributors to it, has several hundred thousand drivers worldwide.[12] In a global labour force of billions that doesn't begin to move the needle. Part-time work increased in importance during the economic crisis of

2008–9, but has ebbed as economic conditions have improved. Still, there is indisputably the opportunity for significant growth in the future.

The question is whether the gig economy will lead to the suspension of the trilemma. The trilemma implies that to scare up enough consumer demand for 'gigs', the price – of the Uber trip or the Task-Rabbit errand, for example – must be low. That, in turn, means that pay must be low. Uber driver wages can't rise to too high a level or Uber will accelerate automation. Similarly, TaskRabbit tasks can't be too expensive, or people will only use the service on rare, higher value occasions, reducing the labour-absorbing power of the service.

A suspension of the trilemma means the arrival of a world of hyper-specialization, in which the market-expanding, match-generating power of the web becomes so powerful that most of the world's billion workers can find themselves a tiny niche that is nonetheless lucrative enough to keep them fed and housed, but which isn't, in the end, doable with software. We can hold out hope for that odd, intriguing world, but we probably should not hold our breath.

The more probable future scenario is one in which new opportunities created by technology – through fracking, or through the disruption of service industries, or through the gig economy – destroy more work than they create, but also reduce the cost of critical goods and services for most consumers. That world has the potential to be a better one, and real standards of living could increase in that world even as pay to workers stagnates.

But realizing that world almost certainly implies a significant evolution in societies' social-safety institutions. As more workers compete for available jobs, wages for people without exceptional skills will stagnate or fall. Eventually, they will fall below what economists call the 'reservation wage': the wage rate at which people decide they are better off not looking for work. Society generally provides an income floor – through welfare programmes and through the support of families and charity. When the available market income falls below that floor, people stop looking for work. They instead choose to live with family, on the dole or state-supplied benefits. That choice will become ever more attractive as technology reduces the cost of critical services and entertainment. A life spent sitting on a sibling's couch watching

Netflix might be pretty miserable, but if the only work available is stultifying and pays very little, then unemployment might nonetheless be the more attractive option.

In a very low-wage world, more people will opt out of work. That will inevitably strain the social-safety net; societies will be ever more clearly divided into those who work and pay for social programmes and those who live off them. Societies will face a reckoning: either they will decide that this dynamic is unavoidable and should be made to work as effectively as possible, or the haves will reduce aid to the have-nots, leading to intense political conflict between those two groups.

That conflict will be shaped and determined by which groups most effectively wield power.

2

Dynamics of the Digital Economy

4

The Virtues of Scarcity

Historically, the labour market's fortunes – as captured in how labour is used within the economy, how it is compensated, and how politically strong it can claim to be – have hinged critically on the extent to which labour is a scarce factor or a plentiful one. When labour is scarce, it can skim off a healthy share of the rewards from economic growth, even if that growth is of an especially technological sort.

In the late 1990s, the San Francisco Bay Area found itself in the midst of an epic economic boom. The world was waking up to the possibilities presented by the internet, and entrepreneurs saw opportunity everywhere they looked. It was obvious to entrepreneurs, to bankers and pundits, and ultimately to all those with an online stock-trading account, that people would use the web to do practically everything; for any economic niche in the brick-and-mortar world, many reckoned, there should be a parallel one in the online world: online banks, music shops, pet stores, universities, and on and on. Being the first to stake a claim to any one of the markets in which the web would transform the competitive dynamics was like buying a licence to print money.

The great dot.com land grab, in which anyone with sense could buy a domain name, crank out a bare-bones business model, take the company public and retire a millionaire (an impressive thing to be in those days), commenced. This was the era of high hopes for companies such as pets.com, the aforementioned online pet-supply retailer, which spent lavishly on advertising before collapsing, and boo.com, an online fashion retailer, which also flopped. But while the hype raged, a more important project was under way: the construction of the hardware and software infrastructure of America's information technology networks, which would persist long after pets.com and its peers had gone belly up. It was firms such as Cisco and Oracle that truly represented the heart of the technology boom.

What was not obvious at the time was just which groups of people would be the big beneficiaries of the technology mania. Would it be big tech investors? The customers using the new technologies? Or the swashbuckling entrepreneurs behind the boom? The answer, as it turned out, was none of the above. The big gains of the era flowed elsewhere; they were captured by participants in the boom in shorter supply than investors, or founders, or customers.

The dot.com mania, as it turned out, was less entrepreneurial than one might have imagined: the rate of entrepreneurship in Silicon Valley was *below* that of the rest of the American economy between 1996 and 2000.[1] Conditions for workers at big firms were simply too cosy at the time to make jumping ship and starting a new company all that attractive, because workers were in desperately short supply.

The unemployment rate in the Bay Area fell to about 2.5 per cent during the peak of the tech boom. Average earnings rose faster in Silicon Valley than elsewhere in California, or America as a whole, and to a level well above that in most other metropolitan areas. On top of that, many salaried employees received part of their compensation in stock options, which were soaring in value at the time. Staying at an existing firm was a highly attractive proposition and so, it is estimated, business creation rates in the Valley were 10 per cent to 20 per cent lower in the 1990s than they were in the rest of the country.[2]

One might describe the economy of the Bay Area at the time by saying that entrepreneurship rates were low because capital was losing out to labour. Capital wasn't scarce – the country and the world were throwing money at the tech economy – but labour was scarcer, as evidenced by the rock-bottom unemployment rate at the time. There was essentially no surplus labour in the Valley, of practically any sort, and especially of the skilled engineers needed to get a tech firm off the ground. To staff a new venture an entrepreneur needed to attract employees away from other companies. That, in turn, meant promising new hires a rather large share of whatever revenues the firm managed to generate – leaving correspondingly less for the entrepreneur himself.

But labour was not the biggest winner of the tech boom. Land was.

The scarcity of labour within the San Francisco Bay Area is hard to understand. Silicon Valley is not a remote fortress; it is part of America's vast domestic labour market. San Francisco's is a pretty good airport, where flights from all over the world routinely land. If firms desperate for workers were throwing money at any able body, one might have expected more able bodies to show up. From 1997 to 2000, average earnings in the region grew by almost 40 per cent: more than twice as fast as earnings across the whole country.[3] And because there were more would-be founders with ideas and financiers with money to spend than there were able workers, firms needed to compete over the available labour, so workers could dictate their terms.

But strangely there was no rush of workers from other parts of the country to take advantage of this fortuitous circumstance. The population of the region didn't explode. On the contrary, Census data actually show a net outflow of residents from the region to other parts of the country during the boom. This seems extraordinary. The Bay Area is a nice place to live. If an engineering graduate even looked at the Bay Area funny, six companies would offer him or her a six-figure salary. And yet people were packing up for other places during the late 1990s.

The repelling force was a scarcity more powerful than the shortage of labour: housing. A worker can't earn a Bay Area salary living in Kansas; he or she has to buy a house that provides access to the Bay Area labour market, but because zoning limits and other regulations make it very difficult to build new homes in the region, the housing stock does not easily stretch to accommodate new workers. Instead, workers who want access to a Bay Area job must bid against other would-be workers for a share of the region's constrained stock of housing. That pushes up housing costs, and fast. At the same time that pay in the region was growing by almost 40 per cent, home prices were nearly doubling.[4]

Workers could extract a huge share of the benefits of the region's growth from firm owners, but landowners could then extract essentially *all* of the gains captured by these same workers. Housing was the scarce factor, and those who owned it were the big beneficiaries of the tech boom.

So, in the Bay Area economy of the late 1990s, labour was exceptionally scarce, and it earned significant rewards as a result. Housing was scarcer, and homeowners did better still. For most of the last generation, however, the world has been awash with labour. Workers have been anything but scarce. The digital revolution is creating an abundance of workers by super-charging automation, globalization and the productivity of a small set of skilled workers.

If the past is any guide, a world in which labour is abundant is not one that is likely to be an especially comfortable one for the providers of that labour. To understand labour's plight, it is important to grasp the economic role of scarcity.[5]

SCARCITY IN ECONOMIC HISTORY AND THOUGHT

Scarcity is one of the fundamental building blocks of economics: economics matters because people cannot have as much of everything as they want, but must accept trade-offs between one scarce item and another. However, economists also realized early on in the industrial revolution that scarcity plays a decisive role in determining which economic participants get the lion's share of the rewards generated by economic growth. New technologies and enterprises boost the total amount of income earned in an economy, but whether that income flows to the inventors of the technologies, the founders of the enterprises, the workers who staff them, or someone else entirely is determined by the relative bargaining power of the players. The group in shortest supply – for whose cooperation everyone else must bid – enjoys a strong negotiating hand.

The Reverend Thomas Malthus[6] was one of a handful of English political economists active in the early nineteenth century who made it their business to build on the writings of Adam Smith and work out the laws of economics. Malthus's working theory of the economy could have been cheerier: he believed the fundamental scarcity of land doomed humanity to misery, and he reasoned that any discovery that boosted agricultural output would simply lead to a rise in population rather than a rise in food (or income) available per person. As growing populations competed for scarce food, war and disease would break out, he explained, thinning humanity's ranks to a more sustainable level. Malthus opposed England's Poor Laws, designed to keep the utterly destitute from dying in the streets; since the poor were doomed at any rate, keeping them alive and capable of breeding simply prolonged and increased their misery, he argued. Happily, Malthus was wrong. Unexpectedly, agricultural productivity grew very rapidly, and families began having fewer children. Malthusian collapse was thankfully averted.

David Ricardo, a contemporary of Malthus, had a more sophisticated take on the relationship between the scarcity of land and the distribution of resources in society. Ricardo was born in London in 1772, one of the seventeen children of a Portuguese family recently relocated to Britain. He made his fortune in finance, making a killing when the price of British government debt soared after news of victory at the Battle of Waterloo (in some versions of the story, Ricardo first encouraged rumours of an English defeat to drive prices down). But he is best known for his fundamental contributions to early economics; he's famous for developing the idea of 'comparative advantage', for instance, which states that trade can make two trading partners better off even if one of them is more productive in every industry. By specializing in the activity at which each is *relatively* best and then trading with the other, each partner profits.

But his insights on the effects of land scarcity are equally significant. Ricardo worried that the owners of scarce land could gobble up most or all of the gains from economic growth, leading to political and economic crisis. He observed that in any society the most productive pieces of land – the ones that could generate the most food for the least effort – were brought into cultivation first, as the residents of a typical village had no desire to work any harder than they needed to provide for themselves. The high-productivity fields provided food at least cost (in manpower and capital), which meant that they could profitably operate when the price of food at the market was very low: that is, when food was abundant.

As the population of a village grew, however, the demand for food rose. Rising demand pushed up prices. And higher prices encouraged landowners to begin farming more difficult plots of land: acreage which took more work and equipment to cultivate, and which, therefore, could only turn a profit for the landowner when food prices were relatively high. This process could continue indefinitely: as populations rose, so too would the demand for food; as demand rose, so too would prices, encouraging farmers to cultivate ever more of the available land. At high enough prices, it began to make sense to adopt really expensive production techniques: such as the construction of levees and dykes to reclaim land from the sea.

Ricardo's key insight, however, was that this process generated an enormous windfall for most landowners. Food prices rose to the point that even the least productive land under cultivation – the rocky, nutrient-deficient, crow-infested, under-watered waste – could make enough money on its harvests to justify planting on the land. But at the price needed to encourage the cultivation of that marginal land, every other landowner profited handsomely. The owner of that first plot of land, the most productive field, continued to produce a bountiful harvest at low cost, and was then able to sell it at fantastic profit, despite the fact that he had done nothing to improve the output of the land or the quality of the crop. The profits flowing to the owners of productive fields were a function of the overall scarcity of arable land.

In a rapidly growing economy in which food supply is constantly under pressure, land is the scarce factor, Ricardo reasoned. As food prices rise to bring more food-growing land online, workers end up handing over a larger and larger share of their paycheque to landowners, who pocket a vast windfall.

This windfall is what economists call a rent. Rent is an economic gain that accrues to someone not because they are doing anything of value, but because they happen to control something scarce that people need. As land becomes ever scarcer, relative to other factors (or, to put it differently, as other factors, such as labour, become abundant relative to land), this factor is able to capture an ever-larger share of national income, thanks to the manufacturers, tailors and publicans spending their earnings on expensive food, lining the landowner's pockets. The more the land bottleneck builds, the more food prices must rise to bring marginal land into cultivation, and the greater the windfall to those who own good, productive land. The worse the land being used elsewhere, the more lucrative it is to be a landowner. This is the paradox of soil; the paradox of productivity.

Back in the early nineteenth century, Ricardo thought that this dynamic spelled doom for European economies. The supply of land was fixed, he noted, while the countries and economies of the day were growing rapidly. Agricultural land would inevitably grow scarcer and scarcer relative to everything else, he reasoned, until landowners were capturing *all* of society's income. Since society clearly could not tolerate that outcome, crisis was unavoidable.

But Ricardo was – at least in some respects – wrong. From 1870, some fifty years after Ricardo's death, to 2005, agricultural productivity soared.[7] During that period agricultural output per person doubled, more or less, despite a quintupling in the global population. This rising agricultural productivity kept a lid on the price of food, spending on which has fallen dramatically as a share of our total consumption. So, too, has employment in agriculture and agriculture's share of the value of economic output.

During that time, too, land ceased to be an economic bottleneck; it lost its fortunate place as the scarce factor in the economy. This evolution was due in part to the expansion of the world's agricultural land. As the world economy grew and became economically integrated, massive breadbaskets such as the American Midwest and the Argentinian Pampas added their output to global markets. At the same time, science and technology enabled fantastic increases in the output that could be generated from a given plot of land, thanks to innovations in capital equipment (including mechanized farm equipment), fertilizer (through nitrogen fixation), and seed stock itself.

Technology and globalization, in other words, conspired to make agricultural land abundant rather than scarce. As land productivity soared, prices fell. And so the windfall to landowners shrank to almost nothing, and the agricultural sector became little more than a footnote in the national accounts, hardly worth noting alongside industry and services. Land abundance means that landowners lack economic bargaining power. When society has more than enough land, some productive fields may sit idle, placing a check on the ability of other landowners to artificially raise food prices by limiting production.

The world market for oil provides another striking example of the way in which the scarce factor in an economy can gobble up an enormous income share. In the 2000s, rapid growth in demand for oil butted up against a relatively fixed oil supply, and oil prices soared. Soaring prices meant an enormous windfall for governments, such as Saudi Arabia, who were sitting on the most productive oil fields. The critical scarcity of oil generated a massive transfer of wealth, from the workers who needed to put petrol in their cars to the oil producers.

Yet globalization and technology soon went to work. New oil fields all over the world came into production. New technologies, such as fracking, massively boosted the supply of retrievable oil in places such as America. As the productivity of oil extraction rose, oil prices fell – as did the surplus captured by the oil producers, and the ability of those producers to wield economic power by manipulating prices.

Yet the nexus between productivity, scarcity and the ability to capture income applies to labour, as well as to land and natural resources.

Imagine that productivity across most workers is extraordinarily high: so high, in fact, that just one person does all of society's labour, providing all the work society needs to have everything everyone wants. One might imagine that this lone worker would be in an economically powerful position. He is providing everything to everyone, after all. But, in fact, this lone worker has almost no economic power. If he limits his production at all in an attempt to extract some rent from the rest of society, then the second most productive worker can immediately step in and capture the whole of the market.

Contrast that world with one in which every last member of society must work tirelessly to meet society's needs. Suppose there is a war on, and the factories must run constantly, using all available labour, to supply the troops with materiel. When everyone, down to the last woman and child, is working, there is no surplus available for owners of the land or machinery. If a worker walks off the job, the factory owner must raise wages until he or she agrees to come back. Economic power, therefore, rests with the workers.

The lessons for workers today are clear. In a world of labour abundance, labour's economic power is pitifully low. Labour therefore finds itself settling for a shrinking share of income – and increasingly irrelevant in the taking of important economic decisions.

FORCES GENERATING LABOUR SCARCITY

What precisely can workers do about their abundance relative to other factors? Labour has occasionally found itself scarce for unique historical reasons. For example, plummeting populations in Europe in the late Middle Ages, a result of the march of the Black Death, significantly reduced the supply of labour relative to agricultural land, leading to a sustained increase in workers' wages. Young America was also a labour-scarce economy with high wages. European-Americans were initially in short supply relative to the land and resources of the vast continent; Native Americans represented a tiny share of the available labour force (a result both of the hostile relations between colonists and indigenous tribes and the massive loss of life among the tribes resulting from initial contact with European explorers).

But the dividend to scarcity has never been any great historical secret, and groups of people have often fought to obtain scarce status for themselves within an economy, at the expense of other groups of workers. Workers seek to make themselves scarce by reducing the capacity of others to compete with them.

The goal of this strategy is to restrict labour supply and therefore affect its price – in the same way a monopolist might profit by cornering the market for oil and only selling a small amount at a time. Supply restrictions work, in part, by diverting some of the gains from production to workers, and away from owners and managers and consumers. Workers benefit by reducing firms' ability to run roughshod over them: using their bargaining power to capture more if not all of the profit generated by production. But if the key to the strategy is limited labour supply, and if supply limits aren't arising as a result of epidemic disease or the challenges of new continents, then artificial labour scarcity will probably only occur through the exclusion of certain groups of would-be workers; historically, labour's power is most often built on the exclusion of outsiders.

This exclusion has often meant discrimination. Virtually every complex society in history has had economic roles that were held off limits to those of the wrong gender, ethnicity, race, nationality, religion or class. Though discrimination can obviously spring from numerous different motives, economic concerns are usually central to any systematic policy of exclusion. The growth of Jim Crow segregation in the American south, for instance, was rooted in a number of concerns, not the least of which was a simple disinclination among white people to have anything to do with black people. Yet white people were also and obviously worried about their economic role in a post-slavery world. Segregation, therefore, ran not simply to matters of where black Americans could sit in public places, but to which kinds of jobs they could hold and how far their educations were allowed to progress. Discrimination was often extremely effective at establishing logistical barriers to competition within labour markets in addition to statutory ones. The systematic undermining of educational opportunities for southern black workers reinforced white belief in their own inherent superiority and created enormous obstacles for these same black workers after the federal government began battling segregation in earnest in the 1960s.[8]

Black people were joined, at various times in American history, by other groups in facing discrimination: Catholics and Jews, Irish and Italians, Chinese and Latin Americans were also excluded in various ways. Within Europe, nationality has discriminated against nationality, denomination against denomination – Protestant against Catholic, Christian against Jew – on and on, going back centuries. The same has always and forever been true of other societies too, from Australia to Argentina, India to China. One hesitates to compare miseries, but one could argue that the most historically egregious of discriminatory tendencies has been that against women. Rules and norms that kept women from full participation in the economy were not simply about dumb bias. They were also about economic power, and the protection of an exclusivity of economic status for men. 'If you compete with us, we shan't marry you,' Alfred Marshall, a nineteenth-century economist with a firmly nineteenth-century habit of mind, once quipped.[9] Economic exclusion is not always central to the narrative used to explain and justify discrimination, but it is practically always a motivation where such discrimination does occur.

Yet the most powerful and durable form of discrimination-induced artificial scarcity is that created by borders. The most extraordinary thing about the American labour market, for instance, is that most of the world's people cannot lawfully participate in it. Over the long run, immigration to rich economies has provided those countries with a source of economic and cultural dynamism, while also (and substantially) raising the standard of living of the migrants themselves. But in the short run, immigrants can be a disruptive economic and cultural force. A particularly large influx of labour can alter the balance of scarcity and influence pay growth: textile workers in the New York City of the early twentieth century had a hard time asking for pay increases, for example, because there were new potential workers getting off the boats every day.[10]

To be very clear, immigration does *not* tend to make workers in destination countries worse off.[11] Over the long run, large immigration flows are a source of economic dynamism: the people who arrive look for and find jobs, but they also spend and invest, create new firms, pay taxes, generate ideas, and contribute to the resiliency and flexibility of the economy – labour markets which repeatedly absorb waves of new arrivals tend to become better at finding economic niches for those same workers.

In the short run, immigration's effects on native workers are not perfectly clear-cut. In some cases, immigration simply substitutes for offshoring and trade, or automation: abundance of one sort substitutes for abundance of another. In other cases, native workers seem to be able to specialize in forms of work requiring more skill and experience that than of the typical immigrant labourer, work which pays better wages. In construction, for instance, native workers might shift into managerial or sub-contracting roles as immigrants do more of the basic labour.

But while some parts of an economy can expand quickly and easily to absorb incoming workers (and respond to the new demand provided by those workers), not all of them can. Cleaning agencies or taxi services can scale up without much ado. Capacity-constrained firms, such as restaurants or construction businesses or factories, take longer. While the process of absorption unfolds, the new labour abundance places downward pressure on wages. More generally, it gives bosses of all sorts an alternative to negotiating with existing workers. And it affects the willingness of existing workers to drive hard bargains: for less-skilled workers, ready work in basic service industries can serve as something of a backstop in the event of unexpected job loss. If those backstop industries are glutted with low-wage immigrant workers, the cost of losing one's better job rises, and one's bargaining power is undercut. Despite the long-run, society-wide benefits to immigration, it is not hard to understand why people of all sorts often favour stiff immigration restrictions.

Not all labour market segregation is so blatantly discriminatory, though. Workers also band together to enforce artificial scarcity by creating guilds, trade unions and professional associations. On the face of things, guild-like institutions look benign enough. They profess to serve as guarantors of professional standards. The American Medical Association, for example, works to make it difficult to become an accredited doctor, and thereby works to raise doctor pay. Yet the public tolerates this because, rightly or wrongly, it sees the barriers to entry erected by the AMA as a way to ensure that doctors are well qualified. Professional organizations can also serve a related function as educational institutions. The European guilds that emerged in the Middle Ages also often set out the professional path for members of a particular occupation, from apprenticeship to master status.

Such organizations, in addition, are seen as important counterweights to owners of other productive factors – either land or capital – who, thanks to the relative scarcity of their contribution to production, enjoy significant bargaining power in negotiations with labour. It is no coincidence that trade unionism emerged as a powerful political force over the course of a tumultuous nineteenth century, a century in which workers often suffered miserable conditions and pay while capitalists prospered.

Collective bargaining addressed the difficulty created by the relative abundance of labour. In the absence of organization, a firm faced little pressure to increase the share of the economic surplus created, when a worker was hired, that flowed to the workers themselves. If the worker didn't like it, he could bugger off, and there would be a long line of replacements waiting to take his spot. Organization sought to eliminate the line of waiting replacements. Firms could have abundant labour at a price negotiated with the trade union's leadership or it could have no labour at all.

But over the last generation, union density in most (though not all) rich economies has fallen steadily, and sometimes sharply.[12] Tumbling rates of unionization are in part a side effect of technological change and globalization, which have shrunk the role of modestly skilled blue-collar workers in an economy, from factory hands to stevedores to printers. Their decline is also owed to political change: to the liberalizing politicians of the 1970s and 1980s who saw the undermining of union power as a way to boost flagging growth. In the years since, the organization of service-sector workers in the US and UK has been a force pushing against the broader decline in private-sector trade unions. Service-sector unions have enjoyed some successes; they have worked, for instance, to support the adoption of higher minimum wages. But trade-union political power is nothing like what it used to be, especially in Anglo-Saxon economies.[13]

A different sort of labour protection, which has multiplied across rich economies, has received comparatively little scrutiny: the occupational licence. Service professionals in an extraordinary range of occupations – including roles such as hairdressing and interior decorating – must obtain a licence to work legally in many states or countries. These licences act as a barrier to entry, helping to protect the scarcity of professionals in a given field, and keep pay higher than it might otherwise be. Licences are often wildly abusive, however. In 2012, a group of Louisiana monks found that they were not allowed to sell the handmade wooden caskets they had been building to earn a little extra money, because they were not a licensed funeral establishment (becoming one would have required them to obtain a casket showroom and embalming room, among other things).* Even if the cost in time and money to get the licence is relatively low, it may deter enough would-be workers (such as those who would normally choose to cut hair a few hours a week, but who can't be bothered to do so when the work can't be done legally without a licence) to prop up the pay of those with a licence.

The more narrow the group of workers involved, the easier it is, politically, to build the necessary exclusive institutions. Groups of professionals have a strong interest in cooperating to lobby for

* The monks took their case to court and eventually prevailed.

certification in their industry: each member enjoys big benefits from the reduction in competition in, say, chiropractic therapy. The higher costs of this certification are distributed over many customers, none of which has a strong interest in devoting the time and effort to a campaign to repeal the certification. And so a thousand professional certifications bloom.

Far better, one could argue, would be a world with many fewer labour organizations taking a much broader view of the welfare of labourers as a whole. Lots of small professional organizations or guilds or unions impede the movement of workers across industries and leave an economy sclerotic and stagnant. Gains achieved by some workers often come at the expense of others, in the form of higher costs for goods and services. One big union, by contrast, has an interest in maximizing the welfare of *all* its members. A broader perspective forces union leaders to think about something closer to the general welfare – to accept that some industries must be allowed to decline while others grow, and that policies which improve economic growth deserve support – while nonetheless exerting bargaining power on behalf of workers.

Trade unions in Scandinavian economies, as well as in Germany, adhere relatively closely to this model. These countries have managed to achieve high income levels with relatively low wage inequality (though top-income inequality in Sweden is surprisingly high). On the other hand, labour share in these countries has fallen, just as it has elsewhere.[14] Ironically, the encompassing labour groups in these economies have periodically embraced wage restraint – that is, keeping wage demands subdued – in order to boost competitiveness relative to their trading partners. In such cases, labour groups work to protect their surplus by siphoning off demand from competitors in other countries.

POLITICAL EFFECTS OF SCARCITY

As the previous section ought to make clear, there is an inevitable political subtext, or even text, to discussions of the economic effects of labour scarcity. Battles over the gains from production are unavoidably political, as is the effort expended by owners of land and capital or by workers to secure the political rights that support scarcity.[15] In *The Wealth of Nations*, Adam Smith mused:

> We rarely hear, it has been said, of the combinations of masters, though frequently of those of workmen. But whoever imagines, upon this account, that masters rarely combine, is as ignorant of the world as of the subject. Masters are always and every where in a sort of tacit, but constant and uniform combination, not to raise the wages of labour above their actual rate ... We seldom, indeed, hear of this combination, because it is the usual, and one may say, the natural state of things which nobody ever hears of ... Such combinations, however, are frequently resisted by a contrary defensive combination of the workmen; who sometimes too, without any provocation of this kind, combine of their own accord to raise the price of their labour ... The masters upon these occasions are just as clamorous upon the other side, and never cease to call aloud for the assistance of the civil magistrate, and the rigorous execution of those laws which have been enacted with so much severity against the combinations of servants, labourers, and journeymen.[16]

If Smith took it for granted that power naturally rests with the master, that might be because he lived during an era of explosive population growth – of labour abundance – in which workers could exercise very little bargaining power within labour markets. They could still organize, or try to, though as Smith notes, employers were quick to call on the government to bring workers to heel. Trade unions were illegal in Britain (and most other countries) until well into the nineteenth century.

Yet workers were not entirely without power. In a series of nineteenth century acts, the British parliament extended the franchise dramatically, until it covered roughly 60 per cent of adult males. The extension of voting rights laid the groundwork for the establishment of the Labour Party, which was a critical force behind the creation of the welfare state in the twentieth century and the adoption of other progressive policies (such as the construction of public housing).

Work by Daron Acemoglu and James Robinson, economists at MIT and Harvard University, respectively, concludes that extending the vote was a rational decision by a political class deeply concerned about the possibility of more dangerous outcomes: including widespread social unrest or revolution.[17] The growth of political radicalism in Britain and the periodic outbreaks of political violence convinced leaders otherwise dead set against the relinquishing of power that such steps could not be avoided if an increasingly mobilized working class was to be sated. The rise of revolutionary or radical politics also contributed to the extension of the franchise in other European economies, including Germany and Sweden.

Workers today labour in a world of labour abundance. They cannot rely on the use of bargaining power within labour markets to capture more of the gains from growth. For the most part, they are also unable to rely on the political heft of powerful labour unions to advance their ends, either in direct bargaining with firms or in political bargaining.

They therefore have little option other than turning to the political system for help. The less succour they receive from existing political institutions, the more open individual workers are likely to be to radical political movements that offer the possibility of political expression and economic power.

The owners of the factors that are scarce, on the other hand, are busy earning enormous fortunes that will persist for years. Tech billionaires and oil magnates, media barons and finance moguls are able to wield market power to accumulate vast wealth, and they will use this wealth to attempt to shape political developments: by supporting ideological movements or financing their own campaigns or donating to candidates.

Because scarcity matters greatly to the distribution of economic rewards, the labour abundance created by the digital revolution cannot help but have significant political consequences.

5

The Firm as an Information-Processing Organism

Most people work for big companies. Across the rich world, about a third of all employment is in firms of 250 people or more, and more than half is in firms of fifty people or more.[1] In rich economies, big firms create a very large share of measured economic value. That means that what happens within firms plays a major role in how work evolves and how economic changes, such as technological shifts, affect workers. Firms are complicated creatures. To a growing extent, they are also social creatures.

In introductory economics courses, the unlucky professors given the job of explaining this dismal science to yawning freshmen inevitably bring up widgets. Widgets are what companies make in the imaginary, simplified worlds economists conjure to illustrate what happens when, for example, the price of widgets rises. Widgets are super easy to make; combine a little labour with a little capital, *et voilà*. One imagines that widget-makers in their widget factories have a pretty clear idea how the business as a whole works, and how their widgeting efforts fit within it.

Real-world economies once bore a much closer resemblance to widget world than they do now. I took my first real, paying job when I was sixteen, developing pictures in the one-hour photo lab of a local pharmacy chain, common in my North Carolina hometown, called Eckerd Drugs. The purpose of that job was refreshingly clear. People wanted to capture memories in photographs, and to have and enjoy those photographs they needed their film developed. The machinery used to do this at the time required a human operator. So there I was, instrumental to a process that added value. The flow of money from customer to store to my paycheque made perfect sense.

Yet, over time (and as one moves further from basic retail or manu-facturing), things become more complicated. In 2004, for instance, I had a different job, as an associate at an economics consulting firm. The terms 'associate' and 'consulting' exude a vagueness of purpose. As an associate, I prepared data and document analyses for senior consultants, who offered expertise to firms or who testified in corpo-rate litigation cases with an economic component (relating to patent infringement, for example). It was nearly impossible to ascertain what work, done by which workers, had made a difference to the outcome of the case. The work we did was information-based, team-produced and hard to monitor.

The bosses at the consulting firm couldn't count up the widgets produced (or photo-lab customers served) and dole out raises or lec-tures at the end of each month. Instead, they constructed workflows and incentive structures to try to nudge team-members towards the sorts of behaviours, such as hard work and cooperation, that gener-ated work that satisfied customers. In the modern economy, the share of activity accounted for by widget-makers is shrinking. The share accounted for by sellers of high-value widget services (such as exper-tise offered to firms to teach them how to use new, networked widgets to maximize productivity, say) is rising.

Workers in the digital era, and especially those working in high-productivity jobs earning good salaries, mostly move information around. Big, successful firms are the ones that structure their internal flows of information in ways that yield things customers want: adver-tising campaigns, trading strategies, productivity-enhancing software, plans to optimize supply chains, and on and on.

What that means, oddly enough, is that the way that information flows within firms is hugely important to a company's performance. The ways that workers talk to each other, or decide what kinds of information to pass along to their bosses, make the difference between success and failure. But that raises a critical issue: when most of a firm's economic value is tied up in the way its workers inter-act, just who should capture the lion's share of the profits when that firm succeeds?

A half-century ago, economic activity was simpler in nature, and the sources of value within a firm were easier to spot. Cars, for example, were simple enough machines that amateur mechanics could tinker with them in their spare time. Automobile manufacturers produced a relatively small number of makes and models, with limited variation. They ordered the parts they needed to do the manufacturing in bulk and carried large inventories. And then they shipped the cars off to dealers to sit on lots for purchase. In the 1980s, this system began to change. Upstart manufacturers such as Toyota adopted lean production techniques, which emphasized close cooperation across all the firms on the supply chain and careful inventory management, as well as constant improvement to both the car designs and the production processes. As electronics shrank and grew more powerful, the operation of the car itself became far more sophisticated; information processing once done within the driver's head was instead handled by on-board computers; variation and personalization took on increasing importance. Today, buyers can customize their vehicle online, and factories can produce completely different models with different features along the same production line (such as the one on the Volvo campus in Gothenburg, Sweden).

A similar evolution played out across many other sectors of the economy. Retail today is about gathering and processing massive amounts of information on customer demand, sourcing products from all over the world, and orchestrating the delivery of those products to shop shelves or doorsteps in a matter of hours rather than days.

As these changes have occurred, the structure of the typical company evolved in response. It has become much leaner. The digital revolution allowed firms to automate or outsource the routine sorts of employment that are easiest to describe and quantify. Back-office work, for instance – managing accounts and keeping tabs on sales and supplies – has been turned over to software in many cases and outsourced to other firms in others. Manufacturing and logistics footprints are likewise shrinking; they are often highly automated or outsourced to supplier firms or both. Digital technology allows companies to turn many straightforward tasks over to machines, while others can be delegated to supplier firms with little risk of loss of quality or control – thanks to the information systems that allow bosses in Palo Alto to keep an eagle eye on production in Guangdong.

The tasks that remain behind – firms' core competencies – boil down to the cognitively demanding work of corporate strategy, product design or engineering. The top carmakers are those who best use software to model vehicle design, to plan manufacture, and to guide the behaviour of the vehicle itself. Sector-leading retailers parse masses of data: about who customers are, what they have bought in the past, what they will want in future, and how products should be marketed, sold and delivered.

Firms are information-processing systems – and, increasingly, that is all that they are. Within the most productive and richest companies, work is increasingly social and cognitive; it is rewarding and well compensated – and open to a relatively small share of an economy's workforce. Even within top firms, a disproportionate share of the value generated flows to ownership and management, who are best positioned to capture the gains produced by employees' interactions. And the concentration of the most valuable bits of the production chain into smaller, highly profitable firms means that workers across the rest of the economy struggle to share in the gains from growth.

Small, brainy companies are responsible for producing enormous economic value in the digital era. The result is a big distributional mess.

THE NATURE OF THE COMPANY

'Why do firms exist?' seems like the sort of question economists should have no trouble answering. Yet when Ronald Coase began probing at the idea in a 1937 academic paper, it quickly became clear that the question was a surprisingly tricky one.[2] Coase was a British economist who lived an extraordinarily long and productive life. He lived to be 102, and still kept busy writing at 100, though his work in the 1930s, when he was in his twenties, was among his most important. It suggested an entire sub-field's worth of mysteries waiting to be understood: a corner of economics now known as industrial organization. Coase won the Nobel Prize for his work in 1991, but his initial question – concerning the purpose of firms – continues to nag at academics today.

In his investigation of how firms work, Coase's starting point was a simple one. People transact in the market to accomplish all sorts of things. When they need milk, they go to the shop to buy it. When they need someone to fix a dishwasher, they ring up a repairman and pay him to figure out what's wrong and fix it. So one might expect the owner of a restaurant, for example, to use similar arrangements to fill the business's various labour needs. A restaurateur might call up independent chefs, servers and bartenders and pay them to complete the necessary tasks, say. The restaurant would be one business, the chef's enterprise another, the maître d's business yet another, and so on.

But that is not the way most restaurants are run. They are not typically set up as single-person enterprises with tens or hundreds of short-term labour contracts with other self-employed individuals. They are instead set up as businesses, which *hire* people to work within the firm's organization. But why?

Coase's answer, which was a good one, was that firms formed when trying to do everything through the market became too big a mess. It takes time and effort for bosses to seek out and hire workers, and for workers to find jobs that best match their skillsets. A restaurateur and a chef sitting down to hammer out a labour contract would need to work out lots of specific details, such as what work of what quality is required for a job to be done satisfactorily, or how the gains from innovation should be divvied up (should the chef use the restaurant's equipment to come up with a new dish, for example), or how much the boss is allowed to interfere with and check up on the chef's work. Employers with trade secrets (the secret sauce in the trademark burger, for instance) risk losing them when contract workers are brought aboard.

Coase reasoned that setting up a firm and hiring people to work for it directly cut down on all of these costs. A firm pays a worker; in exchange, that worker consents to provide their labour within the bureaucratic hierarchy of the firm in question. Within that hierarchy managers need not worry about rebidding a job each time they want to tweak an employee's job responsibilities; instead, they can observe how staff assignments play out and adjust on the fly, safe in the knowledge that workers will do as instructed. A salary, then, is just as much a fee for the worker's obedience as for their labour.

Coase's insight, though important, is incomplete. For one thing, creating a firm doesn't magically eliminate transaction costs. Bosses are not all-knowing and all-powerful, and firms don't suddenly gain the ability to monitor and influence a worker's behaviour by making that person an employee of the firm rather than an independent contractor. Instead, firms have to build an internal incentive structure, which tells employees what behaviours will earn them promotions and bonuses (or get them sacked). Such a structure may be easier to set up, in some cases, than a bunch of contracts with freelancers, but it isn't costless. It requires that management has a clear idea of what it wants the workers to achieve, that they experiment with incentive structures to motivate workers towards that goal, and that they also keep an eye on everything, to make sure the system is working as hoped.

The digital revolution makes it far easier and cheaper to keep an eye on certain sorts of workers and assess their performance. As a result, the boundaries of the typical firm have shifted. Jobs that once needed to be done within a company have been moved outside it. Rather than relying on employed foremen to monitor a production line, motivate or discipline workers, and report back to managers, those managers can simply check the data coming in from a plant in China, engaged on contract, and warn the Chinese contractor that if too many components continue to fall outside specifications their contract will be terminated. Contracts are attractive when the quantity and quality of a worker's output is easy to observe.

Employment is more attractive when the specific contributions a worker makes are hard to measure, for instance because of the highly collaborative nature of the work. In such cases, firms develop incentive structures to encourage behaviour within the workplace, rather than relying on payments in exchange for the meeting of specific production goals. In the digital era, firms increasingly push routine, quantifiable tasks towards contractors, leaving a core business within which work is social, collaborative and guided by broad incentives – by firm culture, one might say.

Incentive structures are, typically, flexible enough to evolve over time. That is: what makes firms work is an evolving internal culture, which turns the firm into something like an organism, struggling for survival in a hostile market environment.

Employers of all sorts value knowledge, and many kinds of knowledge translate right across the economy. Facility with a foreign or software language, for example; the ability to organize data and run statistical regressions; familiarity with petrol engines and a knowledge of how to fix them; or the ability to write clearly and concisely. But much of the economically important knowledge in an economy is 'firm-specific'. A worker's deep knowledge of the proprietary software built and run by one firm will not be entirely able to translate that knowledge to another: while some of the knowledge built up while working on the in-house code may be of use, equally, some will simply not apply to the software used elsewhere.

Other knowledge is even less transferable: an awareness of which workers within a firm are good at which tasks, for example, or how contentious decisions are typically made within a particular firm culture. People who stay with a particular firm for any length of time quickly pick up lots of little, difficult-to-classify pieces of information about how everything fits together to make the place function. Knowledge of this sort builds and evolves over time. Some of it never exists outside the heads of its employees. Some of it is written down: in mission statements or marketing materials or motivational posters in the company loos. Much of it begins life informally and later, if it works, becomes institutionalized in a firm's structure. A team of employees who have been doing a joint task one way for ages may one day decide to do it somewhat differently, then very differently, all of their own accord.

Part of what makes the firm's information-processing machinery work is the knowledge contained within every worker's head: the culture of the firm.

CULTURE AND THE FIRM

Those of us who labour within big firms work within evolved firm cultures every day. When we join a firm we learn all sorts of things about how daily business is done, only some of which may be part of any formal training. We learn how to behave on a daily basis: who to take direction from and how to gather the material needed for a finished work product to, most importantly, what sorts of behaviour will be rewarded. Everyone within a company functions within this environment, and the effect is to detect information, filter it upward to decision-makers, and then generate an active response. That flow of information determines which products get built and how, or which trades are made, or whether a new technology platform decides to focus on growth in users rather than revenue. The decisions that emerge from that flow determine the success or failure of the firm.

Within my workplace, *The Economist*, a weekly publication in operation since 1843, a strong culture has developed. So, too, have a broad array of business practices and a thicket of weekly production rhythms. Some practices persist, whose origins have been lost to time: the steps through which finished pages proceed before being released to printers, for instance, developed over the course of the last century, through a period in which printing technology (and the power of printing unions) changed quite dramatically.

It's not always clear to any of us whether we continue to do things as they have always been done because the procedures had some unanticipated productive benefit that persisted after we switched from older publishing methods to a digital process, or whether they're simply a set of vestigial habits that could, and probably should, be cleared away. *The Economist* is often reluctant to do such ground clearing, and not simply because it is (in its internal governance, at least) a somewhat 'small-c' conservative organization. Rather, the publication's historical success appears to be rooted in the way editorial structures aggregate the dispersed insight and abilities of the journalists and editors into a nice, and profitable, weekly package. It is a process that often seems to function as if by magic. Or as if the editorial workings of the publication operated like a single, efficient information-processing organism.

Some aspects of *The Economist*'s culture are overt. The employment hierarchy is no mystery. The 'hardware' of the firm is also well defined. We have physical offices, including a London headquarters. Other parts of the network are less rigid. *The Economist* has a detailed style guide, for example, itself evolved over decades of publication, which serves as a reference for writers.

Some parts of the culture, including many of the most important, are difficult to describe. It is possible that it is written down somewhere what kinds of stories are meant to go in the finance section of the weekly edition, what the mix ought to be, when the pieces should be filed, and so on. I couldn't tell you where, however, and those involved with the finance section never refer to written directives when thinking about things like that. *The Economist* is what we all understand it to be. The general sense of how things work lives in the heads of long-time employees. That knowledge is absorbed by newer employees over time, through long exposure to the old habits. What our firm is, is not so much a business that produces a weekly magazine, but a way of doing things consisting of an enormous set of processes. You run that programme, and you get a weekly magazine at the end of it.

Employees want job security, to advance, to receive pay rises. Those desires are linked to tangible performance metrics; within *The Economist*, it matters that a writer delivers the expected stories with the expected frequency and with the expected quality. Yet that is not all that matters. Advancement is also about the extent to which a worker *thrives* within a culture. What constitutes *thriving* depends on the culture. In some firms, it may mean buttering up the bosses and working long hours. In others, it may mean the practice of Machiavellian office politics. In others it may be the taking of individual initiative to pitch new ideas or products. At *The Economist, thriving* means lots of things: among them a feeling of collective responsibility for the quality of the articles that leads writers to participate in intense debates over the editorial line, to cooperate selflessly on articles, and to acquiesce to the many layers of editing pieces go through in order to tap the collective wisdom of the staff. Our culture is powerful, and powerfully constraining in some ways, and is the reason the firm produces what it produces.

Any large complex firm will work like this to some degree. It must inevitably rely for its success on the hope that its culture – the 'code' workers follow – gets the right information to the right people at the right time. Indeed, despite the evolutionary nature of many of the firm structures that facilitate this information processing, there is nothing ideal about the system that results. Vestigial connections and routines can be hard to spot and clear away. Local incentives may encourage all sorts of wasteful behaviour. Firms can and do fail, and not simply because competitors are offering clearly superior goods or services.

This culture, or what economists often call 'intangible capital', is increasingly a firm's most important technology. Knowing what information matters and what to do with it is the difference between a wildly profitable company and a bankrupt one.

DARK MATTER AND DISRUPTION

Intangible capital consists of the hard-to-grasp behavioural infrastructure that makes modern firms tick. It rests at the heart of most successful firms, from Apple to Goldman Sachs to Honda, and determines how people work and what sort of salary they are able to earn in return.

Intangible capital includes boring but important stuff such as intellectual property – patents and trademarks – or the value of a widely recognized brand. But it also includes general internal know-how. Firms can invest in intangible capital; indeed, when technology is changing rapidly they must: new technologies create the possibility of doing things far more effectively, but to take advantage of that possibility the firm must learn new ways of doing things. The time required to build that intangible capital accounts for part of the delay we observe between the arrival of a powerful new technology – such as supply-chain management software – and the productivity dividend that technology eventually generates. To use the software well firms needed to hire new workers with complementary skills. They needed to invest in equipment, including computers and scanners, to track inventory. They needed to bring suppliers into the system and train all the workers involved on how to use the new software. Most importantly, they needed to develop internal processes for integrating the new way of doing things with the old culture. Not every culture is as compatible with a new technology as it needs to be to survive; another part of the delay between a technology's arrival and its effect on productivity is attributable to the time it takes for new firms to pop up and drive old ones out of business.

Intangible capital is becoming more important over time. In the 1970s, big firms were tangible animals. A recent analysis considered how much it would cost to duplicate the average firm on the S & P 500: that is, how much you'd have to spend to obtain the machines, buildings, technology, workers and so on that represent the visible components of a company. In the 1970s, the value of those components added up to more than 80 per cent of the firm's valuation. The rest of the valuation constituted what was then defined as 'dark matter': the stuff you can't just go out and buy. Today, however, these proportions of value are reversed. More than 80 per cent of the value of top firms resides in these intangibles – stuff that simply can't easily be accounted for; the buildings and salaries and all the rest of it are only a small chunk of what makes a valuable firm valuable.[3]

This momentous shift occurred as firms shed prosaic operations that could be outsourced to other firms, and concentrated instead on the critical, value-generating work of the business. Half a century ago a major American manufacturer needed to keep its factories onshore, near to the headquarters. Now, however, Apple, a major American manufacturer, can do nearly all of its manufacturing through contractors on different continents: production half a world away, done by other companies, is closely monitored and controlled by technology executives in Cupertino. The Apple that remains (the core, as it were) is both extraordinarily valuable and extraordinarily *intangible* in nature. There is value in the brand and the intellectual property. And there is value in the strange magic that lurks among and within Apple engineers, helping them devise products other firms struggle to emulate. Value in society is increasingly built on ideas, and the firms that do best in this society are those that can manipulate ideas most effectively.

The information-processing role of the firm can help us to understand the phenomenon of 'disruption', in which older businesses struggle to adapt to powerful new technologies or market opportunities. The notion of a 'disruptive' technology was first described in detail by Clayton Christensen, a scholar at Harvard Business School.[4] Disruption is one of the most important ideas in business and management to emerge over the last generation. A disruptive innovation, in Christensen's sense, is one that is initially not very good, in the sense that it does badly on the performance metrics that industry leaders care about, but which then catches on rapidly, wrong-footing older firms and upending the industry.

Christensen explained his idea through the disk-drive industry, which was once dominated by large, 8-inch disks that could hold lots of information and access it very quickly. Both disk-drive makers and their customers initially thought that smaller drives were of little practical use. They were tiny and cheap, it was true, but too slow and with too small a capacity to satisfy users of the bigger drives. Yet the small drives began to find niches – in personal computers, for example, and improved in quality at a phenomenal pace. Customers across the computing industry then switched to the smaller drives en masse, leaving makers of the bigger drives at a loss (in more ways than one).

As Christensen pointed out, this story is remarkably common. IBM was an untouchable behemoth in the mainframe computer market, but it found itself struggling to keep pace with rivals when personal computers stormed into the marketplace. Big-box retail giants caught venerable institutions like Sears completely flat-footed. Digital photography and online photo-sharing apps completely gutted the enormous film-photography industry, and eliminated the jobs of its hapless teenage photolab technicians, in a remarkably short period of time.

The concept of disruption has burrowed deeply into the popular imagination. In a 2014 piece in the *New Yorker,* writer Jill Lepore poked fun at the phenomenon's omnipresence:

> [E]veryone is either disrupting or being disrupted. There are disruption consultants, disruption conferences, and disruption seminars. This fall, the University of Southern California is opening a new program: 'The degree is in disruption,' the university announced. 'Disrupt or be disrupted,' the venture capitalist Josh Linkner warns in a new book, *The Road to Reinvention*, in which he argues that 'fickle consumer trends, friction-free markets, and political unrest,' along with 'dizzying speed, exponential complexity, and mind-numbing technology advances,' mean that the time has come to panic as you've never panicked before.[5]

It does indeed seem entirely straightforward to think of technology as working in this way: enabling start-ups to upend powerful and profitable companies. But it's not obvious that things *should* work this way. Big, profitable firms have lots of money, which could be used to invest in new technologies or buy rivals. They typically have skilled workforces. One might think they would be well positioned to adapt to change.

Yet often they are not, paradoxically, because of the very thing that comprises such a substantial percentage of their value: the 'intangible capital' built into their structures. Company cultures evolve slowly in a way that allows a firm to thrive in a particular competitive setting. Successful companies thrive by picking up market signals – such as what sorts of new features customers would like to see in a product – and reacting to those signals in ways that protect the firm's competitive position. But when customers are offered an alternative solution, wildly different from what has come before, the habits and behaviours and patterns of information flow in existing firms are often poorly prepared to handle the threat.[6]

The news publishing business provides a vivid example of difficulty in adjusting to disruption. Between 2004 and 2014, as we have discussed, newspaper advertising revenue in America fell by more than half. Since 2006, newsroom employment has fallen by a third.[7] Venerable publications like *Newsweek* have gone belly up. Others have been saved from collapse by billionaires willing to prop up loss-making enterprises.

Ad revenue has been battered by the loss of classified advertisements and real estate listings to other websites, such as Craigslist. Yet most titles have also lost subscribers. Readership is down because of an explosion in alternative news sources: upstarts, some of which have built enormous audiences, which are disrupting legacy media.

When journalism on the web first began to appear on the radar of the world's editors and publishers, in the early 2000s, the threat seemed a small one. It was difficult, for one thing, to imagine millions of subscribers leaving behind traditional media to read clunky websites on a desktop computer screen. That seemed especially true given the nature of the content. Early web journalism was often poorly written or edited. It often lacked veracity. There were gaping holes in coverage; and while politics, sports and technology were covered obsessively by early websites and blogs, other important subjects were ignored. When the *Huffington Post* appeared in 2005, packed full with content provided free from crowds of bloggers, it was difficult for readers and editors alike to see the site as a rival to *The Economist, The New York Times* or the *Guardian*. Slicker web publications like *Slate* (which was created by Microsoft in 1996) enjoyed a better professional reputation, but appeared to lack a business model.

Most legacy publications therefore invested in websites, but few focused on the web as the critical marketplace for the future of news.

But the world changed. Readers around the globe became ever more comfortable reading things on screens of all sizes. News aggregators (such as blog readers and Google News) and social networks made it ever easier to find interesting and relevant online content.

Just as importantly, online journalism enjoyed critical advantages over its legacy competitors. One was its price: often, it came for free. There were others: faster response times to developing stories, conversational writing that engaged readers, and a willingness to experiment with new formats. But the most important was a lack of print baggage. New web publications built their internal organizations and incentives – developed their culture – out of nothing. They were free to develop their processes in a way that fit the business of online journalism rather than having to find ways to accommodate a print culture to that new world.

Today, there is no debate about whether digital journalism will threaten print-based media. Every legacy publication recognizes the competitive challenge, but one could imagine a world in which legacy publications, having understood the digital threat, adjusted their strategies and came to dominate the world of web journalism just as they did print. After all, big legacy publications have large staffs full of talented journalists. They have skilled editors. They have foreign bureaus. They have valuable brand names and reputations.

Yet, at the moment, an extraordinary amount of money is being bet on digital start-ups, while legacy publications circulate frantic memos debating how to get their ink-stained veterans thinking digital. Even young journos at old establishments find it hard to think differently; they respond to the incentives they're confronted by and absorb the culture as it is. Simply bringing in tech-savvy millennials isn't enough to kick an organization into the digital present; the code must be rewritten.

Not all economic change is disruptive. Relatively minor changes in the economic landscape may be perfectly comprehensible to the existing structure of an older firm. A change in technology shouldn't wrong-foot an established firm if it doesn't much alter the nature of what is valued in the marketplace. It can, in fact, work out splendidly in some cases. The early decades of the digital revolution were very good ones for many legacy publications: computers allowed firms to shrink production costs dramatically while also improving product appearance. Email made it much easier to communicate with foreign correspondents or gather information from abroad. But more dramatic changes present problems.

At *The Economist*, the challenges presented by online competitors have long been present, and we have developed a set of digital products in response: a website, a suite of blogs, tablet and smartphone versions of the print edition, and a films division. But the same internal structures that make production of the print edition so magically efficient hinder our digital efforts. Everything within our editorial offices is geared towards the creation of a particular product, from the schedule of the work week to our sense of what is newsworthy, to the way we are accustomed to writing pieces: with a given style and with constraints on length and presentation. These structures place limits on our imaginations: we find it more difficult to come up with creative ways of presenting stories online than organizations who have nothing else to think about. They affect our approach to the web in subtle ways: we look in different places for what is interesting than digital publications do; less on social media and more in traditional dailies and in conversations with other print journalists or our sources.

Habits like these affect the resources legacy publications devote to digital work. Research on firms challenged by disruption finds that they tend to invest more in incremental innovation than in projects that go off in radically different directions.[8] *The Economist*, for instance, has placed a high priority on developing a top-quality tablet version of the print edition: a useful digital product to have but one which still relies on the basic print model, in which subscribers pay for a weekly news package (indeed, the very same one print subscribers receive in the mail). We have also invested in more radical approaches, such as the building of a multimedia department. Yet research also suggests that radical investments by established firms tend to be much less productive than similar investments by start-ups, because of the constraints imposed by existing incentive structures.

And that is the real obstacle, both within *The Economist* and at other threatened businesses. Legacy structures are a direct hindrance to innovation, it's true, but their most important effect is in the strong signal sent to workers, that what continues to matter most is the print product. Editors routinely direct journalists to devote more time and thought to online content: to blog posts and multimedia, for instance. But journalists are time-constrained, and priorities must be set. Culture dictates that, when deadlines loom, print comes first. It is hard to develop a best-in-class online product when all the internal signals shout that developing a best-in-class online product should only be attempted after normal print responsibilities have been handled.

Media is hardly the only industry to face these dynamics. Computer makers like IBM developed highly refined structures that made them difficult to beat in the development of mainframe business systems. Those internal structures also left IBM slow to appreciate the threat from desktop machines and clumsy in taking on the desktop market when it finally made its move. Indeed, IBM sought to get around the constraints of its internal mainframe culture by setting up a semi-independent unit to build a PC business; it was eventually forced to 'repatriate' the unit, and effectively concede the market, when workers within the legacy mainframe unit complained of the negative effects of the PC unit on their ability to sell products.[9]

A firm that has spent decades evolving an internal culture optimized to excel in one economic landscape will struggle mightily to adapt that culture to a new environment. All the little incentive structures that kept the worker sub-routines humming in the old world can prove an extraordinary burden in the new one. So while old firms sometimes survive and thrive amid change, upstarts are remarkably successful at exploiting new opportunities, thanks largely to the blank organizational slate they bring to the contest.

THE SOCIAL FIRM

The heart of the firm – the thing that gives it a reason for being – is this cloud of culture and incentive structures, this programme or evolved structure, which transforms the firm into an information-processing machine. A firm's culture helps determine whether it succeeds, fails or limps along. It shapes the firm's response when transformative new technologies appear. One of the critical roles of the entrepreneur is to create space for new firm cultures: new ways of doing things optimized around a technology, flying beneath the radar of more successful firms.

For workers, culture is more personal. It governs the day-to-day environment within which people do their jobs. And it influences the bargaining power they have when negotiating pay. Culture determines how enjoyable or miserable work is, shapes the professional trajectory workers are able to follow.

The purpose of a firm's culture is to encourage behaviour that produces the sorts of business results that bosses like: innovation in some cases, reputational dividends in others, revenue or profits in others. Money matters; workers work for their salaries, and firms will in some cases link pay or bonuses to particular performance goals. (Culture also shapes the particular way in which people approach salary negotiations: in some firms workers are expected to wait patiently for financial rewards, while in others the squeaky wheels get the grease.)

But while money is ultimately the reason most people are there and working, it can't easily be used to shape day-to-day behaviours. Instead, work is collaborative and highly social. And financial incentives are closely linked to opportunities for advancement; workers labour for the right to move up the ladder, to occupy more lucrative positions, or for positions with greater responsibility and freedom. The reward is promotion and pay rises over the long-term, to those who thrive within the culture.

And that is what success within a firm means: learning about and thriving within the culture. In complex firms, where interaction is important and work is not easily quantifiable, discrete work products are not people's main output. Instead, navigating the bureaucracy is an essential part of the job, and the bureaucracy itself is an essential part of the productive process. In crummy firms, bureaucracy means pointless meetings and intolerable firm politics. In healthy firms, bureaucracy has a purpose: sharing information among those who need it, and soliciting ideas for how best to act on that information, for instance.

For workers, work within these sorts of environments means engaging with co-workers and finding a role within the social group. It means learning what sorts of behaviours are favoured or frowned upon by one's supervisor, and what aspects of a finished work product receive the most scrutiny. It means discovering what sorts of information the boss wants and which sorts they would prefer to do without, which decisions they would like left to them and which they would rather be made without them.

With respect to peers and to colleagues senior and junior, it means understanding what serves as a motivational currency. It might be mutual goodwill. Some workers can be motivated to do things out of an altruistic desire to help another (or not to let the other down). Professional respect is another example: appealing to someone's pride in their work can be a useful way to motivate them. For some workers, motivational efforts are more nakedly transactional: they will help if they can expect help in future. Many workers accept tasks or complete them in a particular way out of desire to build and maintain a particular reputation: as the person who is always willing to take extra work, or the person who is preternaturally detail-oriented, or the person who can complete the massive task in no time at all.

But the most successful cultures are often those that promote a true sense of camaraderie. They are places in which workers like and respect their colleagues and help them because cooperating is edifying. They are places in which employees develop a sense of ownership; they work hard because they identify with the firm and want it to do well, both because they take pride in the firm's activity, and because they perceive that an increase in the firm's status also raises the status of the worker. Such cultures are seductive, potent things. Surrounded by talented people working together to produce amazing results, by people who are often friends, and who are engaged in a mission in which one truly believes, one can just about forget that the whole thing is, at some important level, about profit and loss.

Incentive structures are intensely *social*. They are communitarian, in a sense. They represent individuals finding ways to get themselves to work towards a particular common goal, with the idea that there are individual rewards to be had at the end. Good firms make human nature and social interaction part of a sophisticated information-processing mechanism. They are human computers. And while the pecuniary imperative and the discipline of the market represent the power source that keeps the thing running, the *incentive structures* that enable the actual processing are communal rather than monetary.

Seeing firms in this way can help to explain lots of phenomena that might otherwise seem peculiar: such as the remarkable persistence of physical office locations when telecommuting is an available option. Where individual work products are the only thing that matter and are easily assessable, telecommuting is a reasonable option. Yet where social interactions are a key part of the productive process, having people together in an office to bump against each other and swim within the culture is critically important.

The social nature of the twenty-first-century company also determines who has access to the fruits of the business. Much, and possibly most, of the rise in income inequality in America over the last generation or so can be attributed to increases in wage gaps *between* firms rather than *within* firms.[10] As the information-processing capacity of firms has grown in importance – as culture has come to matter more – those working in successful firms have come to enjoy a critical advantage over those working elsewhere.

Yet even within successful firms, the increasing importance of culture has shaped the distribution of rewards. The pay of top executives has risen relative to the firm average right across the economy. Why?

Imagine a firm with a strong internal culture, for instance, and in which that internal culture is a key ingredient in the success the firm enjoys. That culture lives in the heads of all the people working in the firm. It makes everyone within the firm more productive, because of the way it influences social interactions (and, therefore, the flow of information).

That culture is a communal thing. No worker could threaten to take it with him when he left; workers would instead be forced to acknowledge that what they are able to earn within a successful firm culture is perhaps much less than what they could earn elsewhere. Culture doesn't belong to anyone. It can't easily be altered by workers or those in top management. And it drives success only because most workers learn and accept the culture as the way things within the firm ought to be done. Yet, in the end, the profits generated by firms with strong cultures must be divvied up.

Workers, for the most part, lack bargaining power. Their familiarity with the culture will count for nothing at other firms, and any new replacement hire will have a strong incentive to master the firm's culture as quickly as possible. Managers and executives can plausibly demand more of a reward. Culture is easily confused with management initiatives. And skilled managers and executives with relevant experience are generally scarcer than other employees: project managers and associates and journalists, and so on. When profits are higher, they can more easily threaten to go elsewhere, firms worry correspondingly more about replacing them, and so a larger share of the firm's profits can be captured by top management. It is much more costly (and much riskier) to fire and replace a chief executive than it is to sack and replace a worker. Directors and shareholders, faced with the need to determine how the firm's surplus is to be allocated, might find it far more attractive to agree with an executive's pay demands than to sack him in order to make sure that more of the surplus goes to workers, the better to reduce turnover costs.

Of course, sacking and replacing *every* worker is very costly and risky, even when there is lots of excess labour available. It might well threaten the valuable internal culture much more than the replacement of an executive. A firm's cultural capital lives in all its employees; if one quits, it is not threatened; if most do, it is. When labour is organized, it can appropriate the returns of this cultural capital (as it deserves to do). When it isn't, the returns are most easily appropriated by top executives.

But when culture is critical to a firm's success, the biggest beneficiaries are bound to be the owners, who receive residual profits. In most cases, this is a fantastic stroke of luck for the ownership. Some major shareholders, generally the entrepreneurs responsible for creating and building the business from nothing, can plausibly claim to have played an outsize role in forming the culture that drives ongoing firm success. Yet strong cultures inevitably evolve on their own. They are shaped by human responses to challenges, occurring within the social context of the firm. It is not obvious to whom the returns of those cultures *ought* to flow. It is obvious where they wind up, however.

6

Social Capital in the Twenty-First Century

Nestled between the Pennines and the Irish Sea rest some of the major cities of England's old industrial heartland. At their centre sits Manchester, a bustling metropolitan area of more than two million people. Manchester's economy is on the up. In handsome Victorian offices alongside new, glass-enclosed towers, Mancunians go off to work each day in professional services, finance, management and all the rest of it, much like people in London, just over 200 miles to the south-east. 'In once rundown places such as the Northern Quarter, there are now chic eateries and bars', *The Economist* noted in 2013. 'Many old warehouses have become flats or nightclubs. Between 1991 and 2011 the city centre's population increased from a few hundred to over 17,000.'[1]

Two centuries ago, life in Manchester was rather different, as Friedrich Engels noted:

> The cottages are old, dirty, and of the smallest sort, the streets uneven, fallen into ruts and in part without drains or pavement; masses of refuse, offal and sickening filth lie among standing pools in all directions; the atmosphere is poisoned by the effluvia from these, and laden and darkened by the smoke of a dozen tall factory chimneys. A horde of ragged women and children swarm about here, as filthy as the swine that thrive upon the garbage heaps and in the puddles. In short, the whole rookery furnishes such a hateful and repulsive spectacle . . . The race that lives in these ruinous cottages, behind broken windows, mended with oilskin, sprung doors, and rotten doorposts, or in dark, wet cellars, in measureless filth and stench, in this atmosphere penned in as if with a purpose, this race must really have reached the lowest stage of humanity.[2]

Manchester, more than anywhere else, represented the final, awful end stage of capitalism to Engels and his intellectual partner, Karl Marx. It was there that the capitalists' relentless push to increase profits reached an extreme that could not help but bring about its own end.

As the existence of modern Manchester indicates, the inevitable end proved not so inevitable. The average income per person in the city today, adjusted for inflation, is about twenty times what it was in the darkest days of the industrial revolution. Both capital and capitalism were made to work for the common man rather than against him.

Yet now, the relationship between labour and capital is shifting once again, against labour. Since the early 1980s, the share of income flowing to capital, rather than to labour, has risen steadily in economies around the world. A debate rages over the reason for the rise. Some reckon it is down to the plunging cost of powerful digital technologies, which makes it ever more attractive to substitute capital equipment for human workers. Others argue that the rise in capital's share is mostly down the soaring value of property.

In fact, both of these stories are right, and both are symptomatic of a broader phenomenon: the increasing return to *social capital*. Social capital is not a new concept. It has been used for several decades to describe social networks and the sorts of information – including beliefs and values – that flow across them. In the 1990s political scientist Robert Putnam argued that declining social capital in America, as measured by falling participation in social and civic organizations, was responsible for all sorts of ills in American society, from rising crime to alienation.[3]

Social scientists write about social capital in many different ways: some focus on the quantity and quality of connections between individuals, while others try to measure the depth and breadth of things like trust in a society, as a barometer for the quality of social capital. In this section, I'd like to use a particular and specific definition of social capital. I will use the term to refer to contextually dependent know-how, which is valuable when shared by a critical mass of people. The social capital of successful firms is increasingly the most important component of their success: the shared understanding of how the firm does what it does is more valuable than the machines it uses or the patents it holds.

Across societies, in fact, it is the depth of social capital – the social capital per worker, if we could quantify it – that matters most in determining the level, growth and distribution of income. Social capital is unlike industrial capital in many ways. It cannot be seen or traded. It cannot easily be measured, except perhaps as a residual – that which is left after accounting for the measureable stuff. Yet in the way it has transformed relative bargaining power, and in so doing concentrated the benefits of growth in the hands of a few, social capital is very much like its physical counterpart; it is playing an economic role that is analogous to the role of industrial capital two centuries ago. Just as workers' ability to reap significant benefits from the deployment of industrial capital was in doubt for decades, so we should worry that social capital will not, without significant alterations to the current economic system, generate better economic circumstances for most people.

CAPITAL AND SOCIAL CAPITAL

Let's begin by defining some terms. Capital and social capital are different, but I would like to argue that they play a similar role in determining the distribution of rewards within firms. *Capital* is productive wealth. It is buildings and machines and computers and all the things labour uses to create the goods and services people want to buy. It is also the ownership rights to those buildings, machines and companies. Much of the wealth that people hold is in the form of shares of stock – ownership rights to a portion of a firm – which may pay a capital income (in the form of a dividend) or deliver capital gains (if sold at a price higher than the purchase price). When economists grapple with how technological change affects workers, much of what they're thinking about is how the new technology affects the relationship between labour and capital.

As we have seen in previous chapters, the return that each factor earns depends on both productivity and scarcity. Companies hire labour and invest in capital; they set about producing whatever it is they produce, and then they sell what it is they produce to customers. The money earned from sales is used to pay the factors. Some of it goes to wages paid to labour; some of it goes to capital owners outside the firm (if machinery is rented, for instance, or if money has been borrowed to build the firm); some of it goes other places – to the tax authorities, for example. What remains at the end is profits. Profits benefit a firm's owners either directly, if they are paid out, or indirectly, by boosting the value of the ownership stakes in the firm.

If profits are fat, workers might think that they deserve higher wages – that is, a bigger share of the pie. If they have bargaining power, they can get it. Individual workers might have bargaining power because their skillset is scarce in the market: because they can credibly threaten to leave, knowing that the firm is not anxious to go through a costly replacement process which might conclude with a new hire at a higher salary. Workers as a group might have bargaining power because workers as a whole are scarce in the market: if a firm is struggling to fill vacancies because there is very little surplus labour in the economy, then it will raise its wage offers. And workers as a group might also have bargaining power because they are organized, and can credibly threaten to withhold their labour, shutting down all productive activity.

But even if workers are organized, there may be a limit to their bargaining power. Wages might rise above the profit-making level, for example. In that instance, firms could then attempt to raise prices, in order to restore profits, but if the market for the goods on offer is competitive, this may not be possible. Higher wages might instead encourage automation, both to reduce the cost of production and to reduce the broader bargaining power of troublesome workers, though a sufficiently powerful union might also be able to control the pace of adoption of new technologies.

In theory, neither workers nor capital should earn payments wildly different from their relative productivities. In practice, wages and productivities can diverge for long periods of time. It happened for many decades in the initial stages of the industrial revolution. It is happening now. In both cases, capital was key to understanding the divergence.

Social capital is not quite as intuitive a concept as plain old capital. Physical capital – buildings and computers et al – shapes the way people behave at work. Social capital – behavioural patterns that live in our heads – do too.

Economists reckon that growth, and especially the very long-run growth that contributes to wide divergences in living standards across countries, depends on the quality of institutions. Institutions are things such as private-property rights and the rule of law: rules of the game that make possible long-term investment in education or physical capital or intellectual property.

But these institutions are not real things that exist out in the world somewhere. You cannot go to Washington or London and visit the place where the rule of law is kept. Instead, institutions are patterns of behaviour that are observed by individuals in the expectation that others will also observe them. They exist only within our skulls. Societies can and do create organizations of various sorts, which become communities within which members share a particular pattern of behaviour. These communities include everything from institutions of government to churches to for-profit firms to cricket clubs. The set of institutions is the social-capital stock of the society in which they exist.

Social capital is individual knowledge that only has value in particular social contexts. An appreciation for property rights, for instance, is valueless unless it is held within a community of like-minded people. Likewise, an understanding of the culture of a productive firm is only useful within that firm, where that culture governs behaviour. The dependence on a critical mass of minds to function is what distinguishes social capital from human capital.

In both our working and personal lives we are constantly communicating with others, and signalling to them our view of what matters and why. That may seem like an insignificant thing, at least where economies are concerned, yet it is not for no reason that firms pay extraordinary sums of money for office space in central cities. It is to bring people together, so they can be around colleagues, clients and competitors. It is to foster social capital.

But while social capital lives in the heads of the people who make the economy go, its benefits flow disproportionately to the owners of financial capital. That mismatch is a source of significant economic trouble.

THE RISE OF CAPITAL, AND ITS DOMESTICATION

The effect of social capital on modern economies parallels, in important ways, the role that industrial capital played in the nineteenth century. The first few decades of that century were some of the most brutal of the industrial era. Workers poured into manufacturing cities from the countryside and died in droves. Those who survived disease, crime and squalor worked in awful conditions then drank themselves insensate. Though industrialization generated impressive productivity gains in a handful of industries during this period, wages generally failed to keep up with the cost of necessities, meaning that for their trouble most industrial workers were left poorer than they had been before. The stage seemed set for the destruction of the system that had grown up across Western Europe.

Through the first century of the industrial revolution, from 1760 to 1860, the British economy grew and productivity rose. Not by much; economic historians reckon that growth in output per person averaged about 0.3 per cent between 1700 and 1830, before accelerating to just over 1 per cent between 1830 and 1860. *But* sustained growth at even those low rates was a striking departure from most of world history to that point.[4]

Yet throughout this period workers derived little benefit from growth; instead it primarily boosted profits. The rate of profit doubled from the mid-eighteenth century to the mid-nineteenth century,[5] and the share of income captured by owners of capital (as opposed to labourers) rose from about 20 per cent to nearly half. Capitalists earned a large and growing share of national income and amassed enormous fortunes. Marx was not imagining things when he wrote his manifesto in 1848.

Two key economic developments seem to have contributed to this outcome. First, new technologies, such as the machinery being deployed in big new factories, led to a huge increase in output per worker. That, combined with the flow of workers into cities from the countryside, represented a massive rise in the amount of effective labour available to the economy. At the same time, the march of technology substantially boosted the return to capital investment. There were more workers than opportunities to employ them, and more investment opportunities than capital to fund them. The result was high returns to capital and low ones to labour: fat profits.

Marx saw in this dynamic an inevitably antagonistic relationship between labour and capital. The bourgeoisie, in its relentless pursuit to maximize profit, worked constantly to reduce labourers to cogs in the machine. 'Owing to the extensive use of machinery, and to the division of labour, the work of the proletarians has lost all individual character, and, consequently, all charm for the workman,' Marx wrote. 'He becomes an appendage of the machine, and it is only the most simple, most monotonous, and most easily acquired knack, that is required of him.'[6]

At the time, ever more of the manufacturing sector was moving towards a factory model. That was in part due to the economic logic of production with large capital equipment. These machines were often big, power-hungry things running in line with water wheels or steam engines. The capitalists who invested enormous sums in their hulking machines had a great interest in seeing that the machines were not damaged through carelessness, but were manned diligently to the greatest extent possible. Big machinery was therefore a power-ful force behind the migration of workers into centralized plants.[7]

Humanity had to be moulded to fit the demands of industrial eco-nomic structures and the machines that powered them. It should come as little surprise that alienated workers – piled into unpleasant cities and manipulated so as to fit as cogs within massive, impersonal industrial facilities, all in order to earn meagre wages while capital-ists profited – perceived themselves to be the dehumanized playthings of a hostile elite. Neither should it come as a surprise that actual political movements came to reflect industrial social structures. The masses were a force to be mobilized and manipulated, and ideologies competed to command the public most effectively.

Marx reckoned that this system could not be sustained indefinitely, and one of two disasters would eventually bring about its end. Either the capitalists would accumulate so much wealth that the scope to make additional profitable investment would decline to zero; when the pie ceased to grow, capitalists would then turn on each other in a fight for larger slices, leading to the collapse of the system. Or, before this could happen, the accumulation of wealth in the hands of the capitalists would first lead to a revolution by the workers.

In fact, neither occurred (except, eventually, in Russia, where, oddly enough, the revolution preceded industrialization). By the latter third of the nineteenth century, the accumulation of wealth in the hands of capitalists, and the massive investment of those savings into new enterprises and industries, did eventually drive down the return on capital, enough so that the share of income flowing to owners of capital ceased growing, but this did not lead to a war of capitalist against capitalist (not yet, at any rate). Instead, capitalists were content with their constant share of economic output because that economic output kept growing as a result of continued technological progress – an outcome Marx did not anticipate. And so capitalists kept saving and investing and earning handsome returns on those savings, just as the wages earned by labourers grew in line with expanding economic activity. The rising tide didn't wash away inequities, but it kept both capital and labour satisfied enough to hold the revolution at bay. From 1875 until the eve of the First World War, the world's industrialized economies were extraordinarily unequal, but rising living standards for workers kept revolutionary fervour in check.

Yet societies were not exactly living harmoniously, either. As Thomas Piketty notes, in *Capital in the Twenty-First Century*, it took the turmoil of the first half of the twentieth century to undo the inequality that developed in the nineteenth. War, taxation, inflation and economic depression destroyed many of the great fortunes of the industrial era. They ushered in an entirely new state structure, in which extensive taxation was used to fund massive welfare states. And that structure ensured that the rapid economic growth of the first few decades after the Second World War was highly egalitarian in nature.[8]

The taming of capitalism was not a smooth or easy or inevitable process, however. Generations of workers suffered and died, with little to show for their participation in industrialization. Others fought bitterly for political rights and economic bargaining power. Still others fought in world wars that, as much as anything, helped create the conditions for modern, post-war, inclusive capitalism. The egalitarian growth and soaring living standards of the mid-twentieth century were never an inevitability.

By the 1970s, in the rich world, both the share of national income flowing to capital, rather than labour, and the share of labour income flowing to the very rich reached historical nadirs. Stagnation in communist economies was hastening the end for Marx's alternative to capitalism. Meanwhile, Marx's capitalist apocalypse never materialized. But if technological progress helped defuse the volatile political climate of the nineteenth century, capitalism's post-war golden era was not an absolute victory for unbridled markets. On the contrary, its realization was also the result of a social revolution (which created the modern, urban, educated workforce), of three decades of horrible violence that nearly destroyed the rich world, and of the new social contract that emerged in the wake of those decades of horror.

THE RISE AND RISE OF SOCIAL CAPITAL

The economic centrality of social capital is increasing over time, and seems to have been especially important over the last generation. Yet it has always been a feature of economic life. To function in the regimented, cooperative setting of the factory, workers needed to have basic social skills. Joel Mokyr, an economic historian at Northwestern University, has noted that the collapse of cottage industry as a result of competition with factory-based industry changed the structure of the household.[9] Where previously home production could coincide with the nurture and education of children, the change in the locus of work forced the burden of education onto firms and society. That education, he writes, had a particular set of emphases:

> Much of this education, however, was not technical in nature but social and moral. Workers who had always spent their working days in a domestic setting, had to be taught to follow orders, to respect the space and property rights of others, be punctual, docile, and sober. The early industrial capitalists spent a great deal of effort and time in

the social conditioning of their labour force, especially in Sunday schools which were designed to inculcate middle class values and attitudes, so as to make the workers more susceptible to the incentives that the factory needed and to 'train the lower classes in the habits of industry and piety'.[10]

Bosses' efforts to inculcate discipline and deference in workers were an investment in a particular form of social capital. If all workers in a factory setting could be made to behave in a particular cooperation-boosting manner, productivity would rise.

The emergence of the factory also required the development of social interactions in a different and subtler sense. Technological and economic complexities in industrial economies were rooted in fundamental scientific and economic progress. The complexity of the equipment and processes being used in industrial production interacted with the complexity of the organization of the firms themselves to create a hugely complicated economic mess, which placed enormous informational demands on the people running the businesses.

As the complexity of operations rose, firms increasingly relied upon a division of information in addition to a division of labour. For every employee to know every relevant technical and practical detail of the production process would almost certainly be impossible, and the resources wasted trying to make sure everyone knew everything that could be known would be intolerably huge. But just as firms, such as Ford, were able to achieve efficiency gains by breaking processes into many smaller tasks, factories could also benefit by storing the knowledge needed to run a plant in many different places, which is to say people, thereby ensuring that problems could quickly be diagnosed and fixed. The allocation of particular forms of knowledge across workers within a firm, and an awareness of the modes of communication that allow that knowledge to be called upon when needed, represents a critical component of a firm's social capital.

In the industrial age, in other words, human labour meant not simply becoming part of a machine, but also part of the larger cognitive structure of the firm. Factories and firms, as argued in the previous chapter, are large information-processing structures.

Yet a series of momentous economic changes beginning in the 1970s and carrying through the 1980s boosted the economic importance of social capital. Economic liberalization and deregulation contributed to this process. Britain and America reduced tax rates and liberalized, and privatized, government-dominated sectors in the 1980s; other European economies followed in earnest in the 1990s. The long process of trade liberalization that had begun in the post-war decades continued, and was joined by a push to open up cross-border capital flows. Integration of the world economy accelerated, raising the economic return to social organizations capable of managing the more complex economic environment.

At the same time, the digital revolution first registered in a significant way in the public consciousness. Advanced manufacturing techniques were on the rise, leading to the automation of large numbers of jobs in automotive plants, to give but one example. Computers were present in the workplace in a way they had never been before. Telephone calls became cheaper and industry took its first big steps towards the use of mobile phones.

The result was a world that was far more globalized, but also one in which the production and trade of rich economies became 'dematerialized'. But that makes it sound like the boats full of shipping containers crossing the oceans held nothing but vapour. In fact, dematerialization boiled down to the increase in the share of the value of the things being produced that was attributable to services.[11] Cars crossed the ocean, but much of the value of the cars being produced derived from the designers and engineers and coders who made the car run much more efficiently, reliably and safely than it had in the past. The classic example of the phenomenon is the iPod: while components for the iPod were sourced across several countries and final assembly took place in China, most of the value accrued to American firms and workers, and the largest share to Apple itself. Apple did none of the manufacturing, but it did do the design and engineering work. It created the knowledge embodied in the product, which was the most valuable part of it.[12]

The dematerialization of production represents the rise of know-how and the increased importance of knowing what can be done and how it should be done, relative to the doing itself. In a dematerialized economy, information flow is everything. Social capital is the human coding that governs the flow of information.

It can be difficult to distinguish several closely related but fundamentally distinct concepts relevant to work and economic growth. Human capital, for example, is valuable knowledge, accumulated through the investment of personal time and energy, but which is not especially context-dependent: a clear understanding of algebra, say, is useful in many different contexts. Tacit knowledge, meanwhile, is human capital that cannot easily be shared with others: how to juggle or ride a bicycle, for example. Tacit knowledge is useful knowledge that might only be shared through close and repeated contact, but it is not context-dependent. Trade secrets, however, are forms of knowledge kept within certain organizations or firms, but the value of which is also not especially context-dependent: if one firm discovered the code another used to solve a knotty computational problem, it would find that knowledge of use without needing to bring on board the culture of the firm that wrote the useful code.

Social capital, however, is like human capital; it is accumulated by individuals through the investment of personal time and energy. But it is only valuable in particular contexts, within which a critical mass of others share the same social capital. If *The Economist* hires an art designer to help produce the magazine, the designer's facility with image editing software transfers perfectly from their previous place of employment and works whether or not anyone else at *The Economist* understands the software. That's human capital. The awareness of how image design fits into *The Economist*'s production rhythm is only valuable because everyone else at *The Economist* shares similar social capital. That knowledge would not be especially valuable at other publications, nor would it do the designer much good to try to rely on the corresponding social capital from her old employer at her new job.

So, as social capital loomed larger within rich economies, it became clear that firms were not the only context in which social capital took on new salience. Its rise also boosted the fortunes of big cities with lots of skilled workers. By the end of the 1970s, deindustrialization and suburbanization had many of the rich world's great industrial cities on the ropes. Populations were crashing. In 1975, New York

City very nearly went bankrupt. Popular cinema was filled with dystopian visions of the urban future, in which street punks ruled the streets of gutted cities.

But from the 1980s onwards, a turnaround was apparent in some of those very same distressed cities. Big cities that had retained a sizable population of highly skilled individuals began to thrive, and then to boom. New York City's population is now as large as it has ever been. San Francisco is an economic powerhouse. Boston is booming. The London skyline changes dramatically from year to year as new skyscrapers are built.

The world's great, booming cities are thriving thanks to their ability to foster the generation and communication of knowledge. This has always been a part of the logic of cities. In 1890 the great economist Alfred Marshall noted that, within cities, 'The mysteries of the trade become no mysteries; but are as it were in the air, and children learn many of them unconsciously.'[13]

Yet over the last generation, this function has become more important, and more lucrative. That is partly because the value of big ideas has risen, thanks to the expansion of the potential market for them to the globe as a whole. A clever financial product can be marketed all around the world. Useful business software can be sold to firms from Tokyo to Tallinn to Tegucigalpa. Cities are idea-producing places, so, as the value to ideas has risen, cities have prospered.

The idea-producing role of cities has become more important thanks to the increased complexity of the world economy. Management of a business in a global economy is a complicated thing, requiring the collection and processing of information from markets and offices around the world. The information underlying many businesses has also grown in complexity. As technology advances – in finance or computing or biotechnology or anything – the ease with which any one person can become expert in multiple fields declines. Collaboration is therefore necessary whenever expertise in more than one subject is needed to make a project or a business plan work. Very clever people need to be around each other to communicate complex ideas, in order to generate even better ideas. Firms and cities facilitate that process. They provide a context within which social capital can be especially productive.

What does the productive application of social capital look like? It is the development of profitable know-how, more or less. Digital technologies create the potential to do all sorts of new things: to develop new forms of media, to build driverless cars and programme them to zip around city streets, to create machine intelligence capable of deciphering human speech or identifying the images in a picture. We have new capabilities aplenty. What is not obvious is how those capabilities can best be used.

Part of what the productive application of social capital looks like is the community of start-ups and new businesses which experiment with new business models using new technologies to see which of them work. Part of what the productive application of social capital looks like is the operation and management of the firms which succeed in this environment, which have internal cultures that are, for whatever reason, good at absorbing the massive amounts of information zipping around the world, figuring out what bits of that deluge can be ignored and which should be digested, and adjusting their business in profitable ways.

In the early industrial revolution, capital made workers vastly more productive, contributing to an abundance of effective labour. The return on capital was high, because the opportunities to deploy capital productively were plentiful. Investment in capital helped to generate accelerating economic growth, but the benefits of that growth flowed overwhelmingly to profits – to the owners of capital – until the opportunities for productive capital investment had diminished somewhat.

The digital revolution is reprising this history. The clever application of new, digital technology is generating, once again, an abundance of effective labour. The return on social capital is high, because the opportunities to deploy it productively are plentiful.

Here we run into some difficulties, however. Investment in social capital helps to generate growth, but only to the extent that it boosts consumption. In the industrial era, a firm that bought a giant machine tool directly contributed to measured GDP. That counted as investment, which gets included in the national statistical accounts. In the digital era, investment in social capital does not register in the data.

When a group of people comes up with a brilliant business model, that doesn't show up in GDP. When a firm reorganizes itself to take better advantage of digital technology, that doesn't show up in GDP. So that is one difficulty.

Another, more important difficulty is this: the benefits of growth are flowing overwhelmingly to profits, that is, to the owners of capital. But capital and social capital, as we have discussed, are not the same.

Within a firm, ownership of physical capital is typically straightforward: the firm either pays for the use of capital owned by outsiders (to rent a building or the use of server space owned by Amazon, for instance), or it owns the capital used by the firm (such as the desks, computers and intellectual property needed to run the business). Were a firm to decide to shut down and sell off its assets, the workers laid off from the business would typically have no claim to the money received from that sale.

Human capital, on the other hand, belongs to workers. If a firm pays for a worker to get an MBA, or to learn a programming language, and that employee later decides to go elsewhere, the firm can't ask the employee to give back the knowledge he previously obtained. It lives in their head and will continue to benefit the worker in future positions. If a firm shut down and sold off its assets, it couldn't very well try to sell off the coding skills of its workers.

But what about social capital? A firm's culture can only generate value if it is shared by a critical mass of people within that firm. Proclamations issued by bosses are not culture. They only become part of the culture if a sufficient share of the workers at the firm incorporates the substance of the proclamation into their understanding of what the firm is doing, and how they should behave within the firm.

What if a firm tried to sell off its social capital? What would that mean? The firm could try to codify all the kinds of knowledge, interpersonal incentive structures, and patterns of behaviour that make a successful firm tick, write that all down in a manual, then dispatch someone with the manual to try to train up a new firm with a new set of employees with the foreign social capital. But it is often entirely unclear which aspects of culture are useful, or what sorts of social behaviour within firms should count as culture (the rhythms with which employees use the loos no doubt differ from firm to firm, but those differences are probably – probably – not part of the culture). Furthermore, just as a proclamation from a boss can only become culture if it is internalized in a deep way in employees' understanding of how the firm works, the attempt to export culture can only succeed if the would-be recipients reprogramme their behaviours in sufficient numbers. But what will be missing in the new firm is an understanding of why the new culture should be embraced. In the initial firm, there is no why: people who come on board confront the new culture, internalize and succeed, or don't and don't.

Culture is a mass phenomenon that lives in the heads of a critical mass of similarly minded people. It can be exported only by exporting a critical mass of the people who share that culture – meaning enough like-minded people that those on the receiving end of the new, foreign culture obviously have no choice but to adapt and learn. It can be destroyed only by shrinking the culture below the critical mass – meaning below the level at which there is no advantage to be gained by those remaining behind to continue with the old culture rather than adapting to the new.

So could a firm sell off its social capital? Only by transferring a sufficiently large number of people who share the social capital in question. And here the ownership issue comes into sharp relief.

How could a firm transfer, wholesale, a large share of its employees? It could sell itself outright, in which case the owners of the firm would benefit from the transfer, and not the employees. It could negotiate the transfer with the other firm and the workers in question, in which case some of the benefit could presumably be extracted by the employees. Or, if the workers in the firm were organized and could agree to leave en masse, then the workers – within whose heads resides the social capital – could negotiate the sale of the social capital *on their own*, and could themselves capture the benefit of that sale. Alternatively, they could remain where they are and demand compensation for the social capital living in their heads.

Social capital is collective. If the workers who possess the social capital are capable of acting collectively, they can extract much of the return generated by that social capital. If they can't, those with more bargaining power within the firm will capture an outsize share of the benefit: owners and managers.

It is useful to keep social capital in mind when thinking about start-ups: what it is they are trying to accomplish and how they go about it. There are times when a start-up has access to a truly unique technology or business model, so novel and extraordinary that no other firm represents a competitive threat. Those times are rarer than you might expect. Most of the time, lots of people are working on an idea, or on variants on a single theme. Quite often, in new industries, there are initially lots of firms trying to succeed, only a few of which survive. That was true of automobile manufacturers in the early twentieth century, for instance: Detroit was once filled with would-be carmakers trying to figure out how to produce cars profitably. Only in time did the industry come to be dominated by a few very large, very successful manufacturers. Internet firms quite frequently follow a similar pattern. At various points in recent history there have been swarms of would-be online mass retailers, crowds of search engines, and so on. Quite often, only one or two firms have survived in each economic niche.

The winners of these competitions may emerge because they got lucky, or because they had a genius founder. But it is worth noting that the firms that survive are the cultures that survive, and quite often the latter is the cause of the former.

Bridgewater Associates is one of the world's most successful hedge funds: since 1975, when the firm opened, founder Ray Dalio has piled up about $45 billion in net gains. He has also built Bridgewater into a sizable company, which employs 1,500 people and will go on doing business long after Dalio steps down to enjoy his enormous wealth. To maintain continued strong performance while growing in size, a successful company must run on more than the will and direction of a dedicated founder. As a company's operations grow and increase in complexity, the individual oversight of the top executive declines in importance, and the strength of decision-making at other levels of the firm comes to matter more. Everything rests on the flow of information: who is given what information, who is empowered to act on it, and how those actions radiate through the organization. Growing companies, as we've seen, evolve internal cultures – that is, they invest in their social capital – to manage this information flow.

In some cases, such as with Bridgewater, investment in social capital is a highly orchestrated process: Bridgewater is famous for its distinctive culture. The firm operates according to a set of more than 200 principles set out by Dalio. Employees are constantly gathering data on each other, which they record on the iPads they carry with them, and they are trained to embrace a system of ruthless transparency and accountability. A *Wall Street Journal* examination of the firm noted that:

> About 25 per cent of new hires leave Bridgewater within the first eighteen months, but the turnover rate declines after that, according to the firm. Bridgewater is a major recruiter of recent graduates from elite colleges such as Harvard University, Dartmouth College and the Massachusetts Institute of Technology.
>
> Those who stick around embrace Bridgewater's philosophy.[14]

Philosophy, sure. Perhaps more importantly, they accept and embrace what it is to thrive within Bridgewater culture. They understand how information flows around the firm: who is able to communicate which thoughts to whom under what circumstances. The norms that govern interactions between people are what determines the behaviour, as outsiders see it, of the company, and what enables or inhibits its success.

BuzzFeed, a slightly younger company, is building its own unique culture in an effort to dominate the worlds of media, entertainment and advertising. BuzzFeed began life as an offshoot of the *Huffington Post*, begun by HuffPo co-founder Jonah Peretti in the 2000s. In the 2010s, BuzzFeed emerged as the best in breed of a new generation of digital media start-ups. The company specializes in producing just the sort of digital stuff – be that an investigative news story or listicle or viral video – that people are likely to love and share. It uses its ability to produce shareable content to expand its audience across platforms and countries, and to produce advertising that doesn't feel like advertising (and which companies are understandably keen to pay well for).

How does it do it? A *Fast Company* profile explains:

> The company's success is rooted in a dynamic, learning-driven culture; BuzzFeed is a continuous feedback loop where all of its articles

and videos are the input for its sophisticated data operation, which then informs how BuzzFeed creates and distributes the advertising it produces. In a diagram showing how the system works, Peretti synthesized it down to 'data, learning, dollars.'[15]

BuzzFeed is growing rapidly and making money. Its strategy isn't especially mysterious. It is building new systems to help it gather and process data, but those systems don't represent the company's critical advantages. Its critical advantage is the fact that everyone at Buzz-Feed knows what it is to be a BuzzFeed employee: to come in each day and run the BuzzFeed programme and produce BuzzFeed. Peretti and those at the upper echelons of the company, who have helped to build BuzzFeed, are largely responsible for developing the outlines of this culture, and for influencing its contours as the company has grown. Yet as the company grows the culture evolves on its own, shaping the flow of information and affecting the behaviour of BuzzFeed, the institution.

A successful culture, like this one, is a fantastic competitive advantage. If *The Economist* management knew every detail of BuzzFeed's corporate strategy, had unfettered access to its data and technologies, and was determined to build a BuzzFeed clone, it would fail, assuredly. Were it to hire all of BuzzFeed top management, it would almost certainly fail just the same. It might possibly even fail if it bought BuzzFeed outright and tried to operate the company as a subsidiary. The value of BuzzFeed, like the value of *The Economist*, is not simply in what it does, but in the fact that others cannot do it as easily because of the role of culture.

This, it seems to me, is the right way to think about what start-ups are often trying to achieve. Consider the world of new media, for example, which those of us at *The Economist* study with the air of the amateur anthropologist. Vox, for instance, is an online-only general news and culture publication founded by a team of young, talented journalists. (Vox operates, at the time of writing, within Vox Media, a larger media company that owns several different publications.) When Vox debuted, it was able to lay claim to a few strategic assets. One was its nifty content-management system, called Chorus, which gave Vox the ability to present its stories in innovative new ways. Another was its stated approach to journalism, which was to help readers 'understand the news', by providing the necessary background and context to understand whatever new thing had happened in Myanmar, to give an example, or in oil markets.[16] And third, Vox had the talent, credibility and accrued audience of its founders.

None of that a successful publication makes. Lots of publications have nifty publishing platforms, and the very best of them are not all that much better than the clunky systems legacy publications like *The Economist* use. New journalistic approaches are nice, but several centuries of journalism have demonstrated time and again that nothing is more easily ripped off than a format or style that seems to be working at a rival publication. And credibility and audience have half-lives. Vox had a good pitch, and that pitch was good enough to get the publication off the ground, well staffed, and through several funding rounds. But long-run survival depends on more than that. It requires culture. Successful entrepreneurs build cultures that facilitate the success of their ventures.

I don't know if Vox will succeed or not. News is a tough, competitive business. Ad revenue is ever harder to come by, and converting free readers to paid subscriptions is no picnic. The conventional wisdom, which may or may not have been borne out by the time this book is published, is that Vox might well be sold to a much larger media company: one which could benefit from a platform with the Vox brand and the Vox culture. If Vox is sold, it will be the equity owners who reap the direct benefits.

That founders or owners might receive the better part of the return on social capital in a start-up somehow seems just. For one thing, early employees in a start-up are often paid in equity, which means that they have a direct ownership stake in the creation of social capital. For another, founders are building culture out of nothing, or nearly nothing, and often giving everything they have to do it.

At the same time, cultures cannot be built by diktat, no matter how dedicated the founder. Jeff Bezos may be a single-minded, irresistible force, but Amazon culture cannot be sustained without the buy-in of the workers, and once the number of employees grows beyond a close inner circle, culture becomes open-source code, constantly rewritten and edited by the people who live within it.

What is more, those who own equity in a firm are able to capture a share of the returns from social capital even after they leave the company, so long as they maintain their ownership stake. For workers without equity, on the other hand, investment in social capital is more problematic. Because firm-specific culture loses value outside of the firm, workers are in a weaker bargaining position relative to the firm than the executives they work for.

SOCIAL CAPITAL BEYOND THE COMPANY

The firm is not the only critical nexus at which social capital matters. Skilled cities thrive amid the digital revolution because they enable the social capital of firms – they are physical places within which workers can come together to swap culture and ideas. But the most important social-capital community is arguably the nation-state. Real GDP per person in a rich American city, such as Boston (~$76,000), is a lot higher than real GDP per person in a poorer city, such as Jacksonville (~$45,000).[17] A partner at a top American law firm might make ten times in annual salary what a partner at a mediocre firm would earn. But real GDP per person in America is *fifty times* that in Africa's poorest economies and more than four times the global average. A rubbish lawyer in a poor American city can still expect to earn vastly more each year than all but the very elite members of developing economies.

Why? This is a question that has vexed economists for more than a century, and I won't pretend to provide a definitive answer. Instead, I will re-categorize the vague explanations provided by economists under the heading 'social capital' in order to reinforce my point.

Countries can become richer by adding more 'stuff' to the production process. One of the things that happens when an economy like China grows from extreme poverty to something like middle-income status is 'capital deepening', or the application of more capital per worker. But capital deepening runs into diminishing returns: after masses of roads have been paved, fibre optics laid and computers purchased, the addition of still more roads, cables and computers does not contribute much to higher incomes. Instead, the people driving on the roads and using the computers must figure out better ways to use the capital they've got.

That, I would argue, is a process of social-capital deepening. Explanations of rich–poor gaps between countries often focus on variables such as 'technological capability', or the ability to use powerful technologies productively, and 'institutional quality', or the presence of rule of law, secure property rights, functioning markets and so on. Both depend on the critical support of social capital. America has a physical constitution and a body of laws and courts and armed police, but it does not function because Americans are constantly forced by those institutions to abide by the law. Instead, life in America mostly operates according to shared ideas about what is appropriate behaviour for life in America. Sometimes those shared ideas are influenced by the actions of the state (which is itself an institution created to channel the will of the American electorate). Changes to laws and rulings by courts influence our behaviour. People create institutions as receptacles and guarantors of aspects of social capital. But our behaviour is also determined by the signals sent to us through those around us, and through the instruments of culture to which we choose to pay attention.

What is the shared knowledge that is America? It consists of ideas about what sorts of behaviour are appropriate, what sorts are frowned upon, and what action is appropriate when people defect from 'normal' behaviour. It consists of ideas about which formal and informal institutions in society are worthy of trust. It consists of ideas about what sorts of outcomes constitute the 'good life', and what the best routes are to attaining it. It is a shared narrative of history and a conception of who belongs in society and who does not.

All countries (and, indeed, many political entities that are not countries) are built in part on social capital. Britons have an idea about what it means to be British, for example. And some countries have a larger stock of social capital than others. The Scottish might have a clearer idea of what it means to be Scottish than the British do of what it means to be British. Strong social capital is not always of a sort that is conducive to growth. ISIS arguably has relatively strong social-capital underpinnings, albeit of a repugnant, malign sort.

Good government, like sustained democracy, is an emergent property of a strong, healthy social capital. Societies with strong social capital can survive and outlast institutional chaos, such as a crisis of confidence in a government. Societies without complementary social capital will not benefit much from the imposition of new forms of government on them by outsiders.

Hopefully, thinking about society in this way allows us to better understand differences in economic performance. Social capital evolves over long periods of time, lives in the heads of those operating within society (but is often embodied in institutions, such as governments or firms), and influences economic behaviour. Some forms of social capital are growth compatible, others are not. In rich countries, norms and institutions encourage the clever application of new ideas to profitable ends, and innovators can take comfort in the belief that their efforts will be fairly judged in the market, and that any returns they earn will not be unjustly seized by others or the state.

But this conception of society raises two important questions. The first is: how can we invest in more and better social capital? How can we encourage social-capital deepening? As mentioned above, social capital can't easily be exported. There is no good way for America to lend social capital to Guatemala, or indeed to impose it, should it wish. Social-capital-rich countries can merely try to create conditions that encourage the accumulation of healthy social capital in poorer countries. The European Union is a grand effort to do something very much along those lines: to create the incentives in peripheral European states with weaker social capital to invest in and deepen the sorts of social capital that are conducive to openness, the rule of law and free markets. International trade agreements and institutions such as the World Trade Organization are another way in which states actively seek to nurture social-capital deepening in poorer countries. Countries constantly use the geopolitical leverage available to them to try to improve the behaviour of troublesome neighbours, and these efforts occasionally bear fruit. Yet we should also acknowledge that countries are not, on the whole, very good at encouraging social-capital accumulation in others.

Luckily, there is one highly effective way to boost social capital per person: accept people from social-capital poor societies into social-capital rich ones. Social capital is simply information about how to behave. When a person learns the information underlying the social capital of one firm or country, the stock of information in the heads of those already within that firm or society is not depleted. When I joined *The Economist* and began internalizing *Economist* culture, that internalization did not cause colleagues to forget some of what they knew about how *The Economist* works. The most reliable way to deepen the stock of social capital is to allow people to move from low social-capital places to high social-capital places.

Can societies with deep stocks of social capital really accept and assimilate new arrivals without limit, without any erosion or evolution in the social-capital stock? It is in the nature of social capital that it can be altered by anyone operating within society; social capital, it is worth remembering, is simply our internal sense of how things work within particular social groups. The dynamic which matters is: where is this person on the margin between deciding which set of social capital to embrace? If most new arrivals find it in their interest to internalize the new social capital, then the social capital of the assimilating entity will not change very much; those already within society will have little incentive to update their view of how society ought to operate.

THE DOMESTICATION OF SOCIAL CAPITAL

The second key question concerns the distribution of the value generated by social capital. Over the last generation, returns to social capital have disproportionately flowed to those with the greatest bargaining power. That is: top managers, owners of physical and financial capital, and owners of land. Workers, of which there has been an abundance, have not been able to demand much of the growing gains from social capital, despite the fact that this capital lives in their heads.

Just how returns ought to be distributed is not easy to determine; subsequent chapters will grapple with the issue. Yet it is worth keeping in mind industrial history. Marx reckoned workers should rise up and take ownership of capital. Instead, workers settled for access to the means of governance. Political tumult in the 1840s led directly to changes in government in France and Germany, for instance, which included dramatic increases in popular participation: France, for one, briefly enjoyed an early period of universal male suffrage, even though further political chaos soon suspended the policy.[18] And political reform led to changes in economic policy that limited some of the worst aspects of the industrial revolution: by limiting children's working hours, for instance.

Over the century that followed, worker power grew. Workers found an ability to counterbalance the interests of owners of capital, to demand a greater share of the fruits of economic growth and of political power, thanks to an investment in a particular sort of social capital: the labour union. Even before labour unions were legalized across rich economies, the threat of collective action, of a political or even revolutionary nature, encouraged governments to take workers' concerns seriously. Over time unions achieved legitimate political power. Britain elected its first Labour prime minister in 1924.

Industrialized economies also used heavy taxes on the rich to pay for their world wars. And in the decades that followed those wars, the political power of labour led to the construction of expansive welfare states. Workers (in most countries) did not seize the means of production; they were not bashful about taking a healthy share of the returns from production, however.

Coming to the present day, among the manifestations of social capital Robert Putnam cited as in decline in America in the 1990s and 2000s were labour unions. And, indeed, across many rich economies the share of jobs covered by unions shrank over the last generation. That contraction both reflected and exacerbated underlying economic trends. Yet the change in social capital that shaped growth and the returns from it was less an outright decline in its stock than a shift in where and how social capital mattered.

It is perhaps inevitable that when a major technological revolution occurs, which undermines the security of what previously represented the 'commanding heights' of the economy, that the critical locus of social capital shifts from a solidarity centred on achieving an acceptable distribution of the returns to mature industries to a more adaptive, entrepreneurial social capital centred on the profitable use of new technologies within new sorts of businesses. And perhaps it is inevitable that within the firms and cities on the frontier of the technological revolution, the sense of identity that predominates is an aspirational one, a sense of shared mission with colleagues and neighbours who have done best out of the new system. That particular social proclivity might be most conducive to the growth of the economy. Yet the distributional implications are unlikely to be especially egalitarian.

To preview arguments still to come in future chapters, workers may yet decide that the returns from social capital should be shared more broadly across the communities which share social institutions. That might be the route to a more egalitarian distribution of income and wealth, but getting to that point, if it is to be the destination, will require bitter political battles: over the spreading of the social wealth, *and* over just which people count as members of the social community.

3

The Digital Economy Goes Wrong

7

Playgrounds of the 1 per cent

London is the richest city in Europe. Real output per person in central London is nearly four times the average in the European Union, and nearly twice that in Europe's other large, rich metropolitan areas, such as Amsterdam and Paris. Strikingly, London is more than twice as rich as the next richest region within Britain. However one slices it, the city is an extraordinary economic outlier.

The wealth of inner London radiates off the streets and storefronts like heat. Office prices in St James, the neighbourhood in which I work, are among the highest in the world. Art galleries line the streets around the main editorial offices of *The Economist*,* stuffed with works with seven-figure price tags. Around the corner, on Jermyn Street, bespoke shirts and suits can be yours for just an arm and a leg. Nearby there are two places to buy yachts. Maseratis and Bentleys roll through the streets.

This corner of the city is home to royalty: both Buckingham and St James's palaces sit within a stone's throw. But it isn't the Windsors buying up all the £5,000 Grand Cru in my neighbourhood; it's the traders. While the big banks operate in the 'Square Mile' (the historic City of London) or Canary Wharf, hedge funds and private equity shops increasingly locate in the West End. Their presence has ushered what was already an extremely tony area to new levels of toniness.

Tech start-ups, by contrast, once concentrated near 'Silicon Roundabout' – the Old Street area, just north of the City – but are now as likely to be found in Shoreditch, in gritty, hip East London, or south

---

* The current offices, anyway. *The Economist* offices were sold in early 2016 and, as of time of writing, the management and staff were due to be moved to new premises in 2017.

of the Thames: on the South Bank or farther south and west near Wandsworth. That's where I live, in a beautiful neighbourhood I can't really afford, surrounded by hard-charging professionals of all sorts, living the high life in a city that has become a playground for the rich, the quite rich, and the really very rich.

London shares space at the pinnacle of the global economy with just a few other elite cities, among them New York and San Francisco. These cities host the working rich, whose skills and habits mesh perfectly with the technologies and institutions of the digital economy, who are responsible for the creation and management of an enormous share of the rich-world's economic value (and whose earnings are a larger share still of national income). Their productivity contributes to the abundance of less-skilled labour (some of which they re-absorb in their households, as nannies, personal trainers and personal shoppers). Their concentration in rich cities nurtures their careers, turns their neighbourhoods into playgrounds for the elite, and abets the capture of an outsize share of the returns to economic growth. The extraordinary cost of the real estate in these pinnacles of prosperity means that they are effectively inaccessible to most of the labour force: to those not able to earn 1 per cent salaries or not willing to pay huge sums to live in minute apartments in inconvenient neighbourhoods. London, like New York and San Francisco and a handful of other extraordinarily prosperous places, is where the digital economy generates its value, and where that value is channelled to those able to wield sufficient bargaining power.

Such places provide the clearest illustration of the ways in which the digital revolution concentrates economic opportunity on a few, and the challenge facing those interested in achieving a more equitable distribution of that opportunity. Cities are the cosy domains of the rich, and the rich would like to keep it that way.

THE LIFE AND DEATH OF DISTANCE

That a few cities should find themselves in this position represents something of a surprise. In 1997 a journalist at The Economist, Frances Cairncross, published a book titled The Death of Distance.[1]

Her book examined the ways in which the digital revolution was shaping and would continue to shape life and business. Though she seemed to be threatening to kill distance, Cairncross in fact foresaw a world in which distance was safe, happy and very much alive. Technology would, in fact, allow us to *embrace* distance, she predicted. Supply chains would be free to sprawl across the globe, thanks to new and better transport technology. So could business; one could have one's accountants on one continent, and lawyers on another: kept at a safe distance, thanks to information technology. People could sprawl too. The cheaper and easier it became to interact and send information digitally, the less need there would be to disrespect distance by crowding together in cities. Better to find a comfortable place somewhere and let one's data do the commuting.

Nearly two decades later, digital technology is better than ever. One can monitor a production process in a factory half a world away in every detail, while having a video conference with people on each continent. Or one can scrap the production process entirely and *print* whatever thing is needed, from digital specifications that can be whipped around the world at light speed. Yet despite this, we have not embraced distance, as Cairncross supposed we might.

On the contrary, we are actually trying to murder it, with bloodthirsty enthusiasm. We are trying to do everything in one place (or, if necessary, a handful of places); we seem determined to get rid of distance once and for all, by making sure there is none of it between us and everyone else. As I write, the second-tallest building in New York City is a residential tower on Park Avenue, populated by billionaires determined to live on top of one another.

Economic power has nearly always been geographically concentrated. Before the nineteenth century, over the nearly 12,000 years in which humans existed in settled communities, urban populations only very rarely approached populations as large as 1 million; when they did, it was typically in the capital cities of great civilizations, such as ancient Rome or Abbasid Baghdad.

The modern economic era, the industrial era, is a decidedly urban era. Industrializing London was home to 1.35 million people in 1825, making it one of the largest cities ever to have existed. But by 1850 it had added another million people. And on the eve of the First World

War, its population stood at about 7.4 million, a metropolitan colossus.[2] Not long after that, however, its population was surpassed by that of New York. The New York metropolitan area, which itself reached the 1 million person threshold around 1860, was home to more than 15 million people just 100 years later; its population is just over 20 million today and continues to rise.[3]

Technology allowed humanity to live in ever-larger cities; which would be impossible without steel and electricity, to say nothing of modern agriculture. But big cities are not just curious side effects of the industrial revolution. They are a technology in and of themselves, without which we would all be much poorer and less productive.

Cities thrive and grow because of what economists call increasing returns to scale: the larger a city grows the more productive it becomes. Without increasing returns, cities could not get very big: new arrivals would make the city more crowded and unpleasant but wouldn't make the local economy more productive. Living standards would fall and people would eventually say to hell with it and move out. Early in industrial history, these externalities were shaped by a very basic fact: it was extremely expensive to move things over land and not quite as expensive to move things via water. Crowding together by a port maximized access to foreign markets. The crowd of the city attracted newcomers: firms looking for employees and customers, and workers looking for firms keen to hire them. Growth fed on itself.

Yet even at the time, cities provided more subtle support to growth. Large urban economies allow for a high level of specialization, which lifts productivity. A small city might only support a few full-time mechanics, who would therefore need to be generalists, able to tinker on machinery of all sorts. But mediocrity is the cost of generalism: forced to tend to many different machines, the mechanic could not become expert at repairing any one. In a large city, on the other hand, there might be enough big factories to support large numbers of highly specialized mechanics, some of which might only work on one particular sort of printing press or die cutter. Such workers could diagnose and solve more problems, faster.

Specialization plays just as large an economic role today. A small-town lawyer must be a generalist. In big cities, by contrast, entire classes of law firms emerge specializing in particular sorts of corporate law as it applies to particular sectors of the economy. Specialization works on the consumption side as well. Restaurants in small cities cannot be too niche or they will go out of business. Large cities, by contrast, are home to enough people with niche tastes to support a diverse array of cuisines and dining styles. A great diversity of high-quality food becomes an attractive force to would-be migrants to the city, adding to the increasing returns that underlie its growth.

Big cities also provide insurance against rough luck. A journalist working in a dense media market, such as New York, won't have too difficult a time finding a new job if their employer goes out of business. In smaller cities, by contrast, there are fewer media jobs to begin with, and openings come along less often. The interpersonal networks running through productive cities reinforce the capacity of such places to provide insurance.

Most importantly, cities enable the rapid collection, analysis and transportation of information. For much of history cities were an important conduit for the transfer of information of any sort: financiers who wanted access to real-time financial data needed to be holed up in the same coffeehouse or tavern as other financiers. Today, vast quantities of information zip around the world in fractions of a second, reducing the importance of proximity for many kinds of communication.* But cities continue to thrive by enabling the transmission of information that cannot easily be sent in emails or texts: the complex ideas and patterns of productive behaviour that are the foundation of high-value production in the digital era.

The San Francisco Bay Area is perhaps the purest example of a city playing such a role. There, talented engineers, ambitious entrepreneurs and savvy investors participate in thousands of running, intersecting conversations: about which technologies are most promising and which are duds, about how to turn a promising technology into a workable business model, and about how to nurture a new start-up

---

* Though, oddly enough, distance still matters for some sorts of communication of basic financial data. The speeds at which algorithmic high-frequency traders process and act on new information is so incredibly rapid that firms will move office simply to get their servers a few metres closer to an exchange, in order to shave fractions of milliseconds off the time needed to send information through a fibre-optic cable.

into a dominant firm. Young engineers fresh out of Stanford join fledgling start-ups and absorb experience and expertise. Some then partner with colleagues met along the way to found their own firms. Successful tech entrepreneurs participate in venture firms and sit on their boards, providing more advice and assistance. Silicon Valley supports patterns of behaviour – a culture – that cannot easily be replicated elsewhere in the world. At the same time, it supports the circulation of particular forms of know-how, to which outsiders cannot easily gain access.

In 2013, a team of clever Silicon Valley programmers and entrepreneurs launched a product they called Slack. It was a platform for communication within firms, which they had developed for their own use, in the midst of a failed attempt to build an online game. As the team worked on the new product, they quickly discovered its enormous potential: to displace email and other clunky forms of office communication, to replace them with something far more natural and, indeed, fun. Slack is something like a running chat room open to members of particular organizations, divided into channels devoted to particular purposes (from working on one specific piece of code to planning after-work drinks). It is just the sort of digital tool that ought to erode the barriers created by distance. Colleagues all over the world can slip into running conversations, see what's been said, and chime in with their own thoughts.

Yet while Slack is a useful way to coordinate activity across multiple cities, it tends to reinforce rather than erode the benefits of proximity. Slack becomes an extension of rather than a substitute for in-person, in-office conversations. The subtext of Slack exchanges is often lost on those who weren't chatting around the coffee machine a few minutes earlier, or who didn't go out with others for the lunch organized on Slack. Tellingly, Slack itself is headquartered in San Francisco; a few other offices are scattered around the world in major tech hubs. The ranks of top executives and investors in the firm are populated by the founders' past colleagues and collaborators at other Silicon Valley firms. Powerful digital technologies have a way of reinforcing the value of being around other highly skilled, highly productive people.

Analyses of modern urban economies reflect this. Economists Ed

Glaeser and Matthew Resseger find that skilled cities get more productive as they grow, while other places don't.[4] This link, they posit,
seems to be a result of the fact 'that urban density is important
because proximity spreads knowledge, which either makes workers
more skilled or entrepreneurs more productive'. Big, skilled places are
good at *making* workers more productive. Workers in places like Silicon Valley earn a hefty wage premium over similar workers in other
cities, but new arrivals don't get the premium all at once. Instead it
builds over time: evidence that the city is contributing to the knowledge and employability of the workers within it.

Skilled cities have also been the crucible of much of the new sorts
of work created over the last generation. A study of job titles and task
content in America shows that, prior to the 1980s, new occupations
were not especially associated with cognitive sorts of tasks; instead
they tended slightly to favour more routine activities, and cities with
outsized populations of skilled workers were actually relatively slow
to adapt to technological change. But this pattern changed abruptly,
beginning in the 1980s, as computers spread rapidly across the American economy. New occupations suddenly became much more
cognitive in nature and appeared most often in places with large
numbers of college graduates. These cities also became a magnet for
other skilled workers.[5] Over the last generation, places that had lots
of highly educated workers a few decades ago have seen a *rise* in their
share of college graduates, while cities that began with low levels of
educated workers have often seen their share of those with college
degrees stagnate or decline.[6]

The economic importance of two sorts of information (both of
which were discussed in Chapter 6) drives the success of the modern
city. One is tacit knowledge: human capital that cannot easily be
transferred without repeated, personal interactions. Tacit knowledge
includes particular skills – such as how to manage a complex global
business – which can be learned by watching others do their jobs, or
through trial and error, with feedback. It also consists of critical
details about the nature of local technological change. To return to
the evolving media landscape: different companies are using different
approaches to digital publishing, in terms of the technologies used,
the way journalists produce their pieces, and the business models

being followed. People within the industry, who observe and interact with competitors as well as colleagues, develop a memory of why particular decisions were made and how they turned out. This knowledge is valuable for those trying to build better media businesses using new technologies. And while key lessons eventually appear in press coverage, or academic papers or books, the whole useful body of knowledge is informally held, in the minds of those living and working within the community. The knowledge is social, in the sense that it is broadly shared within a particular community. It is not context-dependent, however; an aspiring new-media baron in Seattle would find the information living in the heads of New York media entrepreneurs useful, if it could somehow easily be downloaded and transferred. The value of the information does not shrink by much when it is separated from the community that generated it; it simply isn't very easy to separate.

Social capital, the second sort of information flow nurtured by big cities, is a different story, as we've seen: context-dependent knowledge that shapes the behaviour of people working within particular communities. Cities provide critical support for an economy's social capital by providing the physical setting for social capital within firms. The use and maintenance of *The Economist*'s social capital largely occurs within our London headquarters. The increased importance of within-firm social capital boosts the economic role of the cities that are best positioned to host lots of productive firms.

The metropolitan resurgence is also built on the rising returns of social ties, in terms of both economic opportunity and general life satisfaction. Cities provide a social fabric of overlapping personal networks which link up people across firms and industries in productive ways. In rich cities, rich people have rich friends. These rich friends gather for after-work drinks, enjoy dinner parties or holidays together, chat while waiting together to pick up kids after school or on the sidelines at their children's football match, and generally interact in friendly ways. Networks of the current and aspiring 1 percenters have become richer and more important over the last generation as a result of the rise in assortative mating: skilled, well-paid men are more likely to marry skilled, well-paid women than was once the case. High-powered couples befriend other high-powered couples and hang out in high-powered groups.

This sounds sterile and pernicious. It generally isn't. For the most part, these are people making their way in the world, befriending and coupling with others that they find interesting or funny or nice to be around, and watching friends and neighbours for cues on how to behave: how to structure a social life, how to get one's children into good schools, how to live the 'good life'. The personal economic returns of this life are significant, however. Overlapping networks of friends and professionals facilitate job-changes into plum new openings. They help join up professional partnerships (and couples). They help promote personal ventures, from books to new restaurants to hedge funds. They provide insurance for those within the community who find themselves out of work.

For a particular sort of skilled, high-earning person, elite cities are edifying, lucrative places to be: to achieve professional success, to find interesting friends and lovers, and to build (and perpetuate) the 'good life'. Yet these places are increasingly inaccessible to all but the very rich.

THE GATED CITY

The price of housing in successful cities around the world has soared over the last generation. This enormous increase in house prices was a key contributor to the crash and recession of 2008–9; yet while that dramatic bust temporarily set back prices in many economies, momentum quickly returned. Housing costs in places such as London and San Francisco are again touching new highs, contributing to a broad cost-of-living crisis for many workers.

There is not much mystery to the surge in housing costs. House prices, like the prices for most things, are a function of supply and demand. The demand to live in skilled cities paying growing wages has increased dramatically over the last generation, for obvious reasons. Housing supply, on the other hand, has in most cases failed to keep pace. In some cities it has come nowhere close.

That is certainly not due to lack of interest among builders. As housing costs have risen the spread between the prices new homes command on the market and construction costs has also increased, representing something like a pure profit opportunity for developers. But seizing that opportunity is no easy thing in places like central London or Silicon Valley, for the simple reason that laws and regulations place strict limits on what can be built. Where gaps in the legal strictures can be found and projects actually move forward, Nimbys often rally to apply pressure on local leaders, the better to change zoning rules in order to shrink or kill new developments.

As interest in living in central areas of cities began to recover in the 1980s, the preceding long period of stagnant housing supply became a factor affecting prices. Many central areas had substantial housing vacancies after decades of depopulation, but much of the available housing had deteriorated dramatically and/or was located in undesirable neighbourhoods where crime and poverty remained serious problems. The stock of well-maintained housing in nicer neighbourhoods was very limited indeed, and new demand for urban housing quickly began pushing prices upwards. As that demand built, it soon became clear that these urban centres were unable to accommodate a population boom anything like those they had absorbed in the past.

Since then, the gap between housing costs and construction costs has widened steadily. If housing supply is free to respond to demand, then when the willingness to pay to live in a city rises above construction costs builders build more in order to pocket the spread as profit. If supply can't easily respond, however, then the existing stock of housing must be rationed, using the price mechanism. The cost of housing must rise until enough would-be residents decide the cost of living in the city is no longer worth the benefit. Across the US economy as a whole, housing is about 38 per cent more expensive than it would be if housing supply could easily adjust to demand, according to one recent estimate.[7] In the tightest markets, such as Manhattan and San Francisco, the effect on prices is considerably larger: most of the cost of housing is attributable to the difficulty of building more.

Other rich-world cities actually perform far worse than America at accommodating would-be newcomers. Office space in Frankfurt is six times as costly as it ought to be; in the West End of London office space is roughly nine times more expensive than it would be if build-ers could easily add more square footage. Geography certainly affects property markets: it is not a coincidence that Houston, which is sur-rounded by flat, open terrain, finds it easier to build than San Francisco or New York. But cities can accommodate an arbitrarily high number of residents by building up. New York has been far more willing to allow skyscrapers than London, which is one of the main reasons that London real estate is so much more expensive, relative to hous-ing costs, than space in New York: the New York metropolitan area added more than three times as many new homes in 2015 as the Lon-don metropolitan area.[8] Yet even New York has large swathes of land in which building is constrained by regulation and structure heights are kept to just a few stories. Highly skilled, rich cities are the most aggressive housing-supply regulators. Cities such as Boston, New York, San Francisco and London are home to concentrations of skilled workers in knowledge-intensive industries like finance, tech-nology and media.

Housing-supply limits and soaring housing costs have a dramatic impact on the structure of rich economies. Most notably, the places that have enjoyed the largest increases in productivity and incomes have not experienced similar rises in population. During the indus-trial revolution the economic importance of cities manifested itself in a phenomenal period of rapid population growth. That is not true of the digital revolution.

The population is still growing significantly in places such as New York and London. Yet, in America, population has grown far more rapidly in Sunbelt cities, such as Phoenix and Atlanta, where eco-nomic growth, while robust, has not generated levels of productivity or income anything close to those in Boston or the Bay Area. Indeed, in the 2000s high-wage dynamos Boston, New York, San Jose and San Francisco, and Washington lost about three million people to other cities (net population growth was a result of international migration and natural population increase great enough to offset the outward flow of American households). The ten greatest recipients of net domestic migration, by contrast, absorbed about three million migrants from other American cities. These cities – among them Atlanta and Charlotte, Dallas and Houston – have wage levels that are, on average, about 25 per cent below those in cities from which American households tend to migrate away, and the share of employment in high productivity, high pay jobs in the gaining cities is far lower than in the losing ones.[9]

But households move all the same because of the yawning gap in the cost of housing. A worker moving from San Francisco to Austin will almost certainly take a pay cut. But her housing bill will fall by much more than her salary, leaving the worker with a much greater real income in Texas. A worker moving from Newcastle to London, whether a plumber or banker, might reasonably expect her salary to double after the move. But the cost of her housing is likely to quadruple. To protect their real wages, many rich-world workers opt to stay in or move to relatively low-productivity, low-pay cities. Wages in such places may grow more slowly than they would in New York or London, but if housing costs also grow more slowly, then real pay may nonetheless keep ahead of what it would be elsewhere.

The large-scale, systematic misallocation of labour into low-productivity cities carries huge costs. Recent economic research shows that American output may be as much as 13.5 per cent below the level it otherwise ought to be as a result.[10] In a $16 trillion dollar economy, that is an enormous loss of output every year, equivalent to more than $15,000 for every employed American. Other researchers find that between 1880 and 1980 the incomes in poorer American regions caught up with richer ones, even as poorer Americans tended to move to richer states. Since 1980, however, these trends appear to have stalled.[11]

The more pernicious distributional costs continue to occur within rich cities. Tight supply restrictions, we have seen, translate rising demand to live in a place into rising housing costs. Rising housing costs translate into rising wealth for property owners and rising flows of capital income for landlords. Rising wealth and income from housing assets ought to serve as an inducement to invest in more housing, but of course that is not possible, to any great extent, because of strict regulations on construction. Instead, that wealth represents pure rents, in the economic sense: an economic windfall accruing to those fortunate enough to control a scarce resource. Around 1900, the value of residential housing wealth in Britain was smaller than its

GDP; now it is about three times as large. In America housing wealth as a share of GDP roughly doubled over the same period. Meanwhile, the distribution of housing wealth has become less equal. In the 1960s, the bottom 90 per cent of households controlled more than half of American housing wealth; such households now account for just over 30 per cent of housing wealth, a striking decline. Matt Rognlie, an economist, reckons that soaring housing values are responsible for much of the increased dominance of capital documented by Thomas Piketty; income from housing accounts for 10 per cent of capital income today, up from 3 per cent in 1950.[12]

High housing costs stunt job growth, squeeze wages and productivity across the economy, and channel the gains from what growth does occur to the rich. So how do we account for them?

ZONING AS CLASS WAR

Zoning codes and other housing regulations exist to balance the economic benefits generated when people crowd together and the costs that crowding can sometimes impose. Rules that concentrate density near transport links or specify minimum building standards make cities safer, more prosperous places. But a realistic assessment of the value of zoning rules requires a pragmatic look at the local politics that drive them. Realism forces us to acknowledge that zoning rules are a critical means by which prosperous neighbourhoods and cities protect their exclusivity.

Nimbys are successful because of asymmetries in local power. Everyone in the city benefits when new residents move in; bigger cities support larger local markets with more opportunities for specialization and trade, and new residents enrich the networks that underlie so much of the economic value of modern metropolitan areas. But the economic benefits of a few hundred new arrivals, such as those that might be housed by a new residential tower, are distributed thinly across all those who live and work in the city. The costs are far more concentrated. Those living right around the new tower will deal with the disturbance of construction. They may lose treasured views. Traffic, on roads and local transit, is certain to increase. New construction, especially of smaller, more affordable rental units, could mean the arrival of younger or poorer residents, who might stay late at local bars or pubs generating a disturbing amount of noise, or whose children might take positions at the local school, crowding out others or introducing a different sort of socio-economic background. Perhaps most importantly, new units on the market threaten the value of existing homes. Abundant housing undermines homeowners' ability to capture the value of local economic growth, in the form of rising property prices.[13]

Because costs are concentrated, communities confronted by proposed projects have a strong incentive to cooperate to lobby against them, to get them downsized when possible, and blocked ideally. Neighbourhoods that win historical preservation status for themselves are especially fortunate. Such designations make any significant new construction much more difficult.

Pro-growth residents and developers do occasionally win local political battles, but these victories are often limited to the biggest projects within a city. A grand new development can motivate pro-growth residents to join together, while the billions at stake for builders mean that little lobbying effort is spared. Yet even as grand towers or redevelopments are approved as a result of such lobbying efforts, hundreds of rulings take place elsewhere in the city limiting new construction. Where less is at stake, organized neighbourhoods tend to win. The growth of new towers, though often economically welcome, can often occur alongside a net tightening of housing-supply restrictions.

The individual motivations of the people who engage in Nimby behaviour, or who vote for anti-growth local politics, is beyond my ken. Undoubtedly many believe they are protecting local quality of life at minimal cost to others. Homeowners who worry about their property value are not bad people for doing so. Indeed, one might just be tempted to applaud the behaviour: it does, after all, represent civic activism built on the stock of social capital that makes desirable neighbourhoods in desirable cities such personally and professionally rewarding places to be in the first place.

Yet the outcome of their aggregated behaviour is clear enough and extremely damaging. It represents the protection of wealth and privilege through exclusion. It is landowners asserting a property right to something they do *not* own: the right to say who shall be their neighbour.

Cooperation among residents to oppose new development is supported by the social capital that flourishes in places with overlapping social and professional networks. Those dinner gatherings and launch parties provide venues for communication among those with an interest in protecting a neighbourhood. Perhaps more importantly, they foster a shared sense of purpose or class identity. That shared identity boosts the political effectiveness of the local community: time and energy spent working towards shared purposes represents an investment in local social capital, which both strengthens the community and secures participants' places within it.

Then, when limits on development contribute to soaring housing costs, it is not the well-off professional elite that are displaced, but more marginal households: renters, who get no benefit at all from rising housing costs, or homeowners with lower incomes, who seize the opportunity to cash in and move someplace where their more modest salaries go further. The neighbourhood that remains is one in which the class of actual and aspiring 1 percenters accounts for a much larger share of the local community. The dominance of a particular class in the area increases the sense of shared identity and facilitates a deepening of social capital. Expensive, exclusive cities are the furnace in which a very rich, persistent class of elite professionals is forged

Social capital thrives where it is in the individual interest of those contributing to it to continue their contributions. But while social capital is rewarding in its own right to those who are a part of the community, and further boosts the economic potential of those communities, it also creates a powerful rent-seeking institution: a community that sees its mission, in part, as protecting the wealth of the community by excluding others.

8

Hyperglobalization and the Never-Developing World

So far, this book has dwelt primarily on rich countries. The rich world is a small club, home to about one billion people, or 15 per cent of the global population, but which accounts for about half of global GDP. The future of humanity mostly depends on what happens in the rest of the world's countries, which are home to six billion people, or 85 per cent of global population, and which will account for most (97 per cent) of projected population growth through 2100. The benefits of industrial development bypassed the developing world for long decades, during which the incomes of countries in Europe and North America soared, and when industrialization finally arrived, it occurred incompletely. Unfortunately, the digital revolution is likely to reprise this experience.

Modern industrial history, most of it anyway, is a tale in which the economies of the emerging world fall ever further behind the rich world in terms of income and living standards.[1] The know-how – the social capital – on which rich-world wealth grows, eluded most poor economies over the past two centuries, but for the occasional one-off success story, Japan and South Korea being good examples.

Over the past two decades, that pattern broke down in spectacular fashion as a combination of economic forces, including the digital revolution, integrated billions of new workers into the global economy. Emerging-market workers represent one of the main contributors to the current abundance of labour. Their entry into global labour markets contributed to the rise of a global middle class – and squeezed the incomes of the rich world's less-skilled workers.[2]

But, crucially, this emerging-market boom was *not* built on a broad improvement in institutions; while the boom would not have occurred without economic reform in China and India, neither country, nor emerging markets generally, developed the deep social capital that allowed rich countries to grow at a steady, reliable pace for more than a century. Instead, the emerging world found a route around its social-capital bottleneck. In place of the painstaking process of developing the capacity to turn ideas and know-how into useful and profitable enterprises across a broad swathe of economic activity, the emerging world found itself able to bite off chunks of the activity taking place in richer economies, and in the process captured some of the fruits of their capacity to grow.

Now the era of rapid emerging-market growth is coming to an end. The digital revolution is contributing to the slowdown, and will make it more difficult in future for poor countries to repeat the performance of the past twenty years. Once again, rich economies will enjoy a near-monopoly on the sorts of social capital required to generate a rich-world income.

Slower emerging-market growth carries with it several serious consequences. Perhaps most importantly, it means that billions of people will remain much poorer than they might reasonably have expected to be. Stagnant incomes in poorer countries will exacerbate political tensions in some regions, and will make it more difficult for emerging-economy populations to adapt to difficulties, climate change among them.

Yet that assumes that the rich world cannot do a more effective job of transferring its valuable social capital to those in developing economies. Transferring social capital to poor countries is hard – nearly impossible, history suggests, despite the concerted efforts of rich countries, international organizations and charities of all sorts. But transferring it to individuals is easy enough; it takes little more than allowing people to move into social-capital-rich societies, to participate fully in rich-world social and economic life. But this is a notion to which most people in advanced economies remain extremely hostile.

The question of just how many immigrants to accept from poorer countries is the most important moral question of the twenty-first century. Evidence suggests rich economies will get it badly wrong. Their populations are in effect saying: the poor will learn to become

rich on their own – a painfully slow process that will leave generations worse off than they ought to be – or they will stay poor.

THE STRUGGLE TO CATCH UP

Given the extraordinary economic success of emerging economies
over the last twenty years, it can be easy to lose track of the ground
yet to be made up. Membership of the rich world is a huge boon. It
gets you an income per person, on average and adjusted for local living costs, of about $46,000 a year; America's average income per
person is $56,000, while that of Latvia, the poorest country consid-
ered to be rich by the International Monetary Fund, is $25,000.[3]

In the developing world, by contrast, average income per person is
just $11,000. Average income in China is about $14,000, while that
in the Central African Republic, the poorest poor country, is just
$637 – 1 per cent of the average income in America. Advances in
medicine mean that differences in real living standards between the
rich and poor worlds are smaller than income alone would suggest,
but by any standard the emerging world is a much poorer place than
the rich world, and life is correspondingly more difficult for its people. There are very good reasons that people migrate in their millions,
risking everything, for the chance of better lives in rich countries.

But things are better than they were – vastly so, in fact. An average
real income per person of $11,000 puts the developing world where
America was, in income terms, in the 1940s. In 2000, on the other
hand, average real income in the emerging world was only about
$4,000, equivalent to American real incomes in 1900. And, in 1980,
it was just $1,500, or about where America was in 1830.[4] On average,
anyway, the emerging world compressed about 130 years of development into just over forty. The 'on average' qualifier is worth keeping
in mind, though: even within countries, large income gaps remain. In
China, for instance, incomes in Shenzhen and Shanghai are similar to
those in some rich economies, but as one moves inland the typical
income falls. In parts of China's interior, living standards are on a
par with those in sub-Saharan Africa.[5]

In 2000 about 30 per cent of the population of the emerging world lived on less than $1.25 a day. As of 2015, that figure had fallen to around or below 10 per cent, again depending on just how one estimates living costs. But for that decline, roughly one billion more people would now be living in abject poverty than is currently the case.[6] But for that decline, somewhere between half a billion and a billion more people would now be living in abject poverty than is currently the case. Yet why were poor countries so poor in the first place?

THE ONLY QUESTION THAT MATTERS

A compelling explanation for long-term gaps in growth rates is the holy grail of macroeconomics. Economist Robert Lucas, a Nobel Prize winner,[7] once famously noted, in a paper puzzling over persistent differences in growth rates across countries, that 'once one starts to think about [such things] it is hard to think about anything else'.[8]

Prior to the late 1990s, convergence between poor and rich countries was the exception rather than the rule. America overtook Britain as the world's technological leader in the early twentieth century and never lost its lead. In the middle of the twentieth century, European economies and Japan began to close the gap with America; later a few other Asian economies followed in Japan's immediate wake: Hong Kong, Singapore, South Korea and Taiwan all managed to make the leap to full rich-country status. Yet convergence with rich-world incomes looked a bit like winning the lottery: the pay-off was huge, but the odds were long.

Why should that be the case? Economists have long wrestled with the question and come up with a few possibilities. In a very basic sense, poor countries are poor because they lack capital. There was a time when a lack of financial and industrial capital seemed like the biggest obstacle to development: poor countries were poor because they lacked the means to finance investment in blast furnaces and assembly lines. Yet over the course of the twentieth century, it became clear that countries could develop manufacturing industries without reaching rich-country income levels.

Economists then mused that human capital, or the skill-level of a population, was the critical variable. For countries to climb all the way up the growth ladder, the thinking went, they required populations that could develop and innovate cutting-edge technologies. Yet that too seemed an insufficient explanation for gaps between rich and poor. While countries tend not to get really rich with relatively uneducated populations, there are well-educated poor countries and unimpressively educated rich ones. Perhaps more importantly, highly educated workers in very poor countries become much more productive when they move to rich countries. That suggests there are obstacles within poor countries to the effective application of people's skills.

While physical and human capital clearly play important roles in generating high incomes, social capital is the indispensable factor. Successful countries have good institutions, such as strong and stable governments committed to protecting personal property rights. Social capital supports the evolution and development of growth-boosting institutions, which in turn support the continued accumulation of social capital.

Within rich economies, people understand what constitutes appropriate social, economic and political behaviour. Society can often be counted upon to encourage this behaviour in individuals. Family members, famous sports stars and pop icons can all be counted upon to reinforce the idea that professional success is a good thing. Yet rich societies also create institutions and vest in them the authority to enforce particular behaviours deemed important enough to enjoy the support of the state (certain individual freedoms or property rights, for example). Rich societies also design checks on those institutions to keep them from accumulating too much authority. These institutions are such a vital part of the social capital of rich economies that they are often mistakenly deemed to be the *cause* of growth and wealth. But healthy democracies and market economies cannot be imposed on societies that lack the underlying supportive social capital; they are emergent phenomena in countries with the right sort of social capital.

And so, historically, rich countries tend to stay rich while poor countries tend to stay poor. 'Rich' and 'poor' are stable equilibria. Rich countries become rich by growing at modest rates over very long periods of time. Poor countries enjoy short bursts of growth which tend to end in sharp reversals; very rarely do poor countries sustain rates of growth fast enough for long enough to push them from poor status to rich status.

Arguably, this is because such episodes require supportive social capital, conducive to long-term investments in physical and human capital, and development of the right sort of social capital is hard. Sadly, social scientists lack a satisfying explanation for how it occurs.

HYPERGLOBALIZATION AND THE EMERGING-MARKET GROWTH SPURT

In recent decades it has become easy to dismiss the importance of social capital, as emerging markets of all sorts boomed. But the economic performance of the past twenty years is an extraordinary aberration in modern economic history. Between the end of the Second World War and the late 1990s, relatively few emerging economies were catching up to rich ones at any given point: that is, were enjoying faster growth in real GDP per capita. Those that did caught up at a snail's pace, growing only about 1.5 percentage points faster per year, according to one estimate.[9] From the late 1990s, however, close to 75 per cent of emerging economies experienced catch-up growth and at a scorching pace, growing about 3.3 percentage points faster than rich economies. This was the BRIC era; in 2001 Goldman Sachs economist Jim O'Neill identified the big emerging markets, Brazil, Russia, India and China, as countries likely to reshape the global economy and financial markets, thanks to their extremely rapid growth.[10] Yet the growth acceleration extended right across most of the developing world.

What happened? The answer seems simple enough: China happened. In 1980, GDP per person in China, adjusted for local living costs, was 2.5 per cent of that in America. By 2015, that figure had risen to 25 per cent, and China had become the world's largest economy. China drove convergence across the emerging world in a few ways. Its rapid growth generated explosive demand for commodities, from copper to oil to rice; it's economic ascent tugged along commodity-exporting emerging markets in its wake; and it invested massively in commodity-exporting countries, largely in infrastructure but also in other areas, to help secure the flow of resources.

China's growth also established it as a 'mega-trader', in the language of economists at the Peterson Institute for International Economics: meaning that trade is critically important to the Chinese economy (accounting for nearly half of Chinese GDP) and that Chinese trade is critically important to the global economy, accounting for more than 10 per cent of global merchandise exports. Mega-trader China has become the hub of Factory Asia: it hoovers up imports from around the region, some for domestic consumption, but an enormous share for processing into exported goods. The growth of trade networks centred on China constituted another mechanism by which its rise boosted the fortunes of the emerging world as a whole.

China was not the only engine of global growth: India, the world's other billion-person country, has also enjoyed an impressive expansion over the last two decades. Yet the effects of China's rise dwarfed those of its neighbour: its boom was longer, stronger, and much more import-intensive than India's, and it therefore had much longer coattails.

Zeroing in on China as the driving force behind emerging-market growth only takes us so far. We are then confronted by a follow-up question: precisely how did China manage it?

One possibility is that, like Japan and South Korea before it, China's institutions evolved in a way that encouraged the accumulation of capital and technological know-how. This is a difficult thing to assess: decades of communist rule have warped the social capital across Chinese civic society. It is clear, however, that in the late 1970s Deng Xiaoping's Communist Party began experimenting with tolerance of market activity and openness to foreign trade. Property rights in China have never been secure, nor has the market been the primary force allocating capital. But property rights have been secure enough to satisfy lots of multinational firms, who have been willing to contract with Chinese companies or invest directly in the Chinese economy.

Yet the role of foreign capital points to a second force at work in China's rise, without which Chinese liberalization would have generated far more meagre returns. Over the last generation, technological change enabled explosive growth in global supply chains. Supply-chain trade has had far-reaching consequences for global development.

Success in export markets once required economies to develop an entire suite of capabilities. To export electronics or cars, South Korea and Japan needed to build an entire, high-quality supply chain domestically: they needed lots of firms capable of designing and manufacturing components, and well-organized corporations capable of planning and coordinating the design, production and sale of complex goods. That took time. It began with countries building and mastering the entirety of supply chains producing relatively simple goods, such as toys and radios, then building on those capacities and expanding, slowly, slowly, into more complicated products: computers, cars and industrial machinery. It was part of a fundamental transformation of the domestic economy to rich, industrialized status.

But supply-chain trade changed everything. A California technology company could source component supplies from half a dozen Asian economies, have them all meet together in a Chinese port city for assembly, and then ship the finished package to consumers. This allowed production chains that previously needed to be located within a single firm or country to fragment across an economic archipelago.[11] Information technology was not solely responsible for these developments: better shipping technologies and trade liberalization helped. Yet without the ability to coordinate production efficiently and in real time, the system could never have developed.

Its effects were profound. Emerging economies no longer needed to slowly and painfully accumulate knowledge and capabilities as they worked their way from production of plastic toys to industrial robots. A country like China could instead immediately get into the advanced electronics export game simply by tapping into global supply chains. Cheap labour and a relatively small set of competencies were suddenly sufficient to participate in production of advanced goods. Trade swelled as international supply chains developed: shipments between suppliers that would not previously have registered as exports increasingly did. And countries that found their way into supply chains enjoyed rapid growth.

Supply-chain trade benefitted emerging economies around the world. In the Americas and in Europe, regional clusters developed in which components made in some countries were shipped to others for assembly and final export to consumers. But the emergence of 'Factory Asia' was the most fundamental effect of the supply-chain revolution. The hyperglobalization that resulted shunted hundreds of millions of low-wage workers into direct competition with less-skilled workers in the rich world and elevated China to its status as the world's largest economy.

The emerging world now represents roughly half of all global output. Developing economies that were once at the mercy of rich-world crises and business cycles can now themselves cast a great economic shadow on advanced economies, or pull them along towards prosperity. Meanwhile, the distribution of global income has fundamentally changed. Prior to the 2000s, global income followed a bimodal, or two-peaked, distribution, with lots of people in the rich world clumped together around high incomes and lots (and lots) of people in the developing world clumped together around low incomes. Now there is something like a global middle class, and a graph of the global income distribution is just one big hump, with many people earning moderate incomes while a small share of the global population earns very high incomes.

THE DIGITAL DIVERGENCE

The great emerging-market boom is now over. In 2015, emerging markets grew at their slowest pace since 2001 (excepting the global-recession year of 2009). The pace of catch-up with American income levels, in terms of GDP per person, has slowed to practically nothing. The proximate cause is the inevitable slowing of the Chinese economy. China's boom peaked in 2007, when the economy notched up an extraordinary GDP growth rate of more than 14 per cent. It grew at less than half that pace in 2015. More declines are inevitable.

The closer an economy gets to the technological frontier, the more difficult it is to achieve rapid progress towards that frontier. At the same time, China's institutions remain highly illiberal. This was a mild hindrance when the order of the day was attracting foreign capital and building modern infrastructure. As China's economy becomes more like those of the rich world, it increasingly faces rich-world sorts of questions: growth becomes about knowing what can be done and what usefully should be done with new technologies. State capitalism may be poorly suited to such decisions.

Slowing growth in China gutted commodity markets, leading to difficult times for the commodity-exporting countries that did so well from the late 1990s to the early 2010s. Meanwhile, trade growth has slowed dramatically. That is partly due to the exhaustion of big gains from supply-chain trade. The growth of snaking production chains across countries supercharged trade growth, since the production that once occurred entirely within one country began to require multiple rounds of exporting and importing. Yet once such chains are in place, trade growth necessarily slows unless chains continue to fracture into additional links or new kinds of products are built along global supply chains. The world has arrived at the point at which neither is occurring at a meaningful pace.

But something else is going on as well. The digital revolution, which helped to establish the supply-chain revolution in the first place, continues to shape trade patterns and the ways in which trade enables development. This time, new technology seems to be making life harder for the emerging world.

Supply-chain-powered development represented an accelerated – if somewhat superficial – form of industrialization. It seems to have also, as a side effect, accelerated deindustrialization. Readers in rich economies will be well aware of the phenomenon – the loss of manufacturing work to other locations – that hollowed out once-great cities like Detroit. Britain, the first industrializer, was the first to face this particular ill, quite early in the twentieth century. Over time, the bug affected more industries in more corners of the rich world: in America, for instance, manufacturing employment peaked as a share of total employment in the early 1940s and declined at a remarkably steady rate thereafter; but there have been particularly nasty spells of employment loss along the way – in the early 1980s, for instance (when Reagan and Thatcher earned the ire of many blue-collar workers) and then in the 2000s. Remarkably, manufacturing now accounts for less than 10 per cent of American employment.

In the emerging world, deindustrialization is occurring at ever earlier stages of development: an ailment that economist Dani Rodrik has labelled 'premature deindustrialization'.[12] When manufacturing's share of total value added in the South Korean economy peaked in 1988, real income per person in South Korea was about $10,000, or just less than half the American level at the time. When that same peak was reached in Indonesia in 2002, its real income per person was roughly $6,000, or about 15 per cent of the American level. And when India reached that point in 2008, its real income per person was only about $3,000, or about 6 per cent of the American level of income at that time.[13] Indeed, Arvind Subramanian, an economist and chief economic adviser to the Indian government, reckons that the Indian experience actually represents something like premature *non-industrialization*, or the fizzling out of industrialization before it ever really got going.[14]

This is extremely worrying. Historically, successful economic development virtually always meant industrialization. It is not clear whether there is an alternative strategy.

Supply-chain trade, which allows low-wage economies to manufacture goods without building the broad set of capabilities once associated with industrialization, leaves poorer countries vulnerable to the premature loss of industry as wages rise. But the increasing dematerialization of economic activity described in Chapter 6 is also undercutting the industry-based approach to development that was the closest thing to a reliable ticket out of poverty in the era before hyperglobalization.

The value in the goods and services we trade and consume is increasingly derived from the knowledge used to create or provide them, rather than the material or capital equipment or labour used in their production. This is easiest to see in the consumption of digital products: the value of an album by one's favourite artist has always derived, in large part, from the creativity of the musicians and the clever marketing of the studios, but in the past that value shared a significant amount of space with the expense of the recording and editing equipment used to generate the album, with the material and equipment needed to produce physical copies of the album (as records, cassettes or CDs), and the time and expense associated with bringing those physical albums to physical locations to be sold. Music today is very different. High-quality recording and editing software can be had for a song, so to speak, and once a complete digital version of a song is complete, it can instantly be transmitted to anyone anywhere in the world. Value in music production today is now *overwhelmingly* about the skill and marketability of the artists, which is almost entirely intangible. That's dematerialization.

This is increasingly true of the physical goods we consume as well. The production of an automobile is still a very resource- and capital-intensive process: you need a lot of material to get it done and a lot of giant machines to stick all that material together. Even so, the value of an automobile is ever more associated with non-physical inputs. Most automobile manufacturers now do the vast majority of their design work – of the cars they produce and of the production plants themselves – virtually, using high-powered design software. Much of the operation of the vehicles is managed by on-board computers, which keep tight control over engine function, assist in actual handling of the car and in navigation, and ease the management of on-board climate control and entertainment experiences. Therefore, much of the labour at automobile manufacturers now consists of designers and engineers, accountants and marketers sitting behind computers, rather than technicians operating on an assembly line. And much of the value in a car is in the sophisticated electronics within it.

Countries still compete for the factories in which the vehicles are assembled: such factories still mean jobs, if fewer than in the past, and jobs are useful things to have in an economy. Yet, from a value perspective, factory assembly is a drop in the bucket. Very nearly anyone can do it. It is no surprise that state governments compete to offer incentives to car firms looking to open new production plants: firms can shop around, and capture more of the value of production, because they are in possession of the scarce know-how needed to make a car – the design and programming knowledge, the capability to manage global supply chains, and so on – while the locations competing for the plant are largely interchangeable.

The story is very much the same for something like an iPhone: Apple captures the lion's share of the return from making them despite its outsourcing of virtually the whole of the production chain because it is the creative force behind the product design. Indeed, it is true of our consumption in general; we once devoted most of our household budgets to physical things: food and drink, clothing and furniture. Now we spend vast amounts on things like education and healthcare, or on housing, the value of which is mostly dependent on the access it provides to social capital rather than the wood in the walls and the plastic in the pipes.

Subramanian describes this shift as one from 'stuff to fluff', and it is reflected in the trade data. If one measures trade in gross terms, by totting up the price of all the things sold across borders, then physical goods are as dominant as ever, accounting for about 80 per cent of trade or roughly the same as the share a generation ago. If one instead measures trade in value-added terms, then shipments of physical goods have tumbled in importance, dropping from about 71 per cent of world exports in 1980 to 57 per cent in 2008. Services are accounting for ever more of the value traded across borders. And trade in 'knowledge-intensive' goods and services (those in which research and development spending or skilled labour generate most of the value added) now accounts for about half of the value of all trade in goods, services and finance.[15]

Developing economies are discovering that this evolution presents them with serious difficulties. The growing importance of knowledge (and the growing irrelevance of other cost sources) means that the advantage to rich-world firms of moving anything abroad is decreasing. 'Reshoring' in manufacturing, or the relocation of industrial production back to the rich economies that were priced out of such businesses decades ago, is often framed as a labour-cost phenomenon and a potential boon for middle-skill workers in advanced economies: with Chinese wages rising, some believe, it is increasingly attractive for firms to keep assembly in America, and to employ thousands of manufacturing workers in the process. But that is not, for the most part, what is occurring. Reshoring is predominantly a function of the rising knowledge-intensity of production, which means that variations in the cost of unskilled labour no longer matter all that much. Better for Tesla to keep production close at hand (in Fremont, California, on the eastern shore of San Francisco Bay) where its skilled engineers can keep a watchful eye on the code operating the plants, than to move assembly abroad in search of modest savings on the wage bill. And sure enough, the reshoring phenomenon, where it has occurred, has not brought back mass employment of less-skilled workers.

That means that economies which were hoping to establish an industrial foothold for themselves by using their low labour costs to wiggle onto a supply chain are increasingly out of luck. There are exceptions, but they are of a particular and unhelpful sort: where labour is so incredibly cheap that it remains economical to use people *in place of* available technologies. But in these cases the advantage to firms of locating in poor economies is precisely that the use of more sophisticated technologies is not necessary, which means that any transfer of technological knowledge to the local workers will be extremely limited, and the rungs which might otherwise have led to a more productive, sophisticated state of economic activity have been removed.

ON THE OUTSIDE LOOKING IN, AGAIN

One might have hoped that the extraordinary growth of the last two decades represented a new normal: the sharp slowdown of recent years suggests it did not. The world might not return to a state in which rich-world incomes grow higher and higher relative to those in poor countries, yet we should not be surprised to discover that the world is still a place in which it is very difficult to become rich, and rare that countries accomplish the leap. Indeed, not even China, the star of the last growth generation, has got there. While incomes in Chinese coastal cities are similar to those in poorer rich-world cities, those in China's vast hinterland, as mentioned, remain very low by global standards.

If there has been a new model of development to come out of the era of emerging-market growth, it might be one in which small pockets within developing economies build the social and technological capacity to compete in the knowledge-intensive global economy. India, which has managed to create a few clusters of technological innovation, provides an example. The growing Indian economy, over 1.2 billion people strong, is a mouth-watering target for online retailers. Serving India's retail needs, however, will require the construction of a massive, sub-continent-wide logistics network, including scores of enormous warehouses. Those warehouses are potentially a source of vast amounts of employment for less-skilled Indians (of which there are hundreds of millions). Yet the falling cost of simple robotics and the increasing power of computing means that many of those jobs may never be created. Instead a very small number of highly skilled Indian programmers may earn a good living writing code to control the robots who travel the great aisles within these warehouses, moving around goods shipments that might otherwise have been handled by human workers.

But these pockets of wealth in poor countries are unlikely to prove sustainable. Governments in those countries will be sorely tempted to grab a large share of the riches, or will be co-opted into corrupt corporatism – into supporting the winners and suppressing competitors in exchange for favours of various sorts – in a way that stifles the vibrancy of the high-value sector, undermining its long-term survival. The temptation for the most successful individuals within such clusters will inevitably be to move to places where the social capital is more supportive of long-term wealth creation.

Getting rich is not about growing fast. Developing economies often grow fast. But then they stop, and when they stop they often perform very poorly. Brazil, which as some observers sourly note is the country of the future and always will be, grew at a blistering pace from 1967 to 1980, at about 5.2 per cent per year. It seemed a very good bet to join the ranks of the rich world. But from 1980 to 2002 average growth was effectively nil: good periods were entirely offset by nasty downturns. The emerging market wave picked up the Brazilian economy once more in the 2000s, leading to a new burst of enthusiasm.

But Brazil is now back to its old tricks, struggling through economic hardship.[16]

Becoming rich, and staying rich, is about consistency. It is about stringing together long periods of modest, positive-on-average growth; about achieving a social state in which long-term technological progress is consistent with political stability. Constant innovation over centuries is an amazing feat, without which we could not enjoy the living standards we do. More impressive, in a way, is the social capital within rich economies, which has given rise to the institutional flexibility needed to manage two centuries of wrenching, dramatic economic development. The process has obviously not been perfectly smooth, but who in the early nineteenth century, if told of the technological changes to come, the effects they would have on the demand for labour, the changes to human life they would necessitate, would bet that democratic governments would continue to persist for long periods of time *and* that they would oversee extraordinarily steady, stable growth of about 2 per cent per year.

The sort of social capital needed to support centuries of sustained growth is extremely difficult to cultivate. The progress of a few fortunate decades can too easily be swept away by a few years of trouble. But the right sort of social capital can be made to cover an ever-larger share of humanity if more of humanity is allowed into the places that have that right sort of social capital. It is worth the time and effort of residents of poor countries to invest in rich-country social capital, if they can relocate to places where that social capital is shared by a critical mass of the population. Advanced economies cannot turn poor countries into rich ones, and we lack a foolproof recipe for poor countries seeking to make themselves rich. What can be achieved, and has reliably been achieved, is the process of helping residents of poor countries to become rich by welcoming them into places with strong social capital.

Mass immigration has always been the obvious, pie-in-the-sky solution to wide gaps in incomes across countries. Yet the experience of the last two decades has left the people of the rich world deeply ambivalent, if not outright hostile, to the notion of increased immigration. Years of stagnant wages punctuated by the trauma of the financial crisis have voters turning inward, looking to fringe politicians of a nativist bent.

Advanced economy institutions, while capable of sustaining long periods of economic growth, have struggled to maintain public support for discomfiting change in an age of hyperglobalization and rapid technological change. And now, even as growth in poorer economies slows in worrying fashion, advanced economies, and the globalized economy itself, are facing retrenchment.

9

The Scourge of Secular Stagnation

In the autumn of 2008, the world economy very nearly collapsed in on itself. In September, Lehman Brothers, a large, globally connected bank, went bankrupt. In the weeks after, shares in most of America's other large banks sank precipitously, creating the possibility of the failure of most of the country's major financial institutions. Lehman's bankruptcy created havoc in money markets – a key part of the country's financial infrastructure that large corporations often use to fund themselves – pushing the economy towards a frightening situation in which corporate giants such as General Electric might have been unable to pay their workers. Only massive intervention by the federal government and the Federal Reserve prevented an economic catastrophe. Even so, the world economy shrank in 2009. Millions of people were tossed out of work. Lives were destroyed. The Great Recession pushed the eurozone into its own existential financial crisis. Now, years later, labour markets remain scarred, and voters are falling into the arms of fringe politicians.

Such severe economic crises are rare. They require a perfect storm of enabling circumstances: lax financial regulation, large-scale capital flows and major policy mistakes, among other factors. But the Great Recession would not have occurred without large economic imbalances, which made management of rich economies difficult for policy-makers. These imbalances are in large part the result of the forces unleashed by the digital revolution. They have left the global economy especially crisis-prone: while the dry tinder that supported the financial conflagration of 2008–9 has largely burned away, the underlying vulnerabilities remain, and seem certain to generate future, costly crisis episodes.

The problem, which rich economies have not come close to solving, is that the gains from growth are not flowing to workers. Instead, they are piling up in the paycheques and portfolios of the rich. Economies do not work very well when purchasing power does not flow to those who want to, and indeed who need to, spend. Until markets, or governments, find better ways to spread the benefits of growth broadly, the world faces the risk of recurring, severe crises.

Economic downturns such as the Great Recession are periods of weak demand. Demand in an economy is the amount of money spent – on everything from cars to computers to trips to the dentist. When demand is weak, too little money is spent to utilize all of the productive capacity of an economy. Firms lose sales and sack workers. In classical economic models such episodes ought not occur: firms should recognize the fall in demand and respond by reducing wages and prices, so that the spending which does take place goes further, utilizing the full productive capacity of the economy and preventing the need for layoffs. In practice, this doesn't work. Prices and wages don't adjust very easily. Moreover, reductions in wages and prices, when they occur, affect people's expectations of how their incomes will grow in future. That, in turn, can put a chill on spending and investing, deepening the downturn.

Much of the rich world, and a surprising share of the emerging world as well, appears to be descending into an era of chronically weak demand. This condition, which economists label 'secular stagnation', is associated with limp and vulnerable economic expansions, which often conclude in the deflation of big asset-price bubbles, and with protracted and disappointing recoveries. Secular stagnation in part of the world can function as a sort of economic black hole, sucking other economies into the weak-demand trap. It is caused by and exacerbates the inequities generated by the digital revolution.

Secular stagnation slowly undermines support for the existing economic order, and while it is possible that governments will eventually settle on benign solutions to the problem, it is more likely that prolonged secular stagnation will lead to a broad backlash against global economic integration, and a costly turn inward.

THE HOARDERS

The idea of secular stagnation dates to the 1930s, when Alvin Hansen, an American economist of Keynesian intellectual disposition, wrote a book called *Full Recovery or Stagnation*.[1] The book mused on the nature of the Depression and asked whether some of the factors behind it might lead to permanent, structural economic malaise. Hansen suggested that ageing populations and a slowdown in technological progress reduced the appetite for investment. With too few profitable investment opportunities available to absorb society's savings, demand would flounder and the economy would slip into stagnation.

Hansen was wrong, as it turned out. Military mobilization pushed rich economies to operate at their fullest capacity, and after the Second World War, governments embarked on massive public investment schemes while households, buoyed by strong wage growth, went on a consumption binge. Yet some economists are now turning to Hansen's stagnation hypothesis as a way to understand current economic woes.

The problem is a disconnect between what is earned in an economy and what is spent. In modern economic life, one person's spending is another's income. You buy a subscription to *The Economist*; I earn a paycheque. I buy a new pair of blue jeans; others – a clothing-store manager, some textile workers in Bangladesh, a designer in New York – each receive a share of what I paid, and go on to spend their earnings elsewhere. What is spent in an economy is what is earned.

Some of what is earned is saved, however. Some small portion of my paycheque is deposited in a bank savings account, while another modest portion goes into a retirement account, where it is used to buy stocks. Those savings pull money out of the economy; they are a drain on demand. But in a normal, healthy economy that drain does not translate into a recession because the savings are recycled into investment. My bank can lend against my savings to a business looking to invest in new equipment; that spending on equipment adds to demand. Similarly, when I buy stocks, stock prices rise, making it more attractive for firms to raise money by tapping equity markets, which can then be used to fund demand-creating investments.

Sometimes, this process hits a bump in the road. A spate of bad news can lead to broad economic pessimism. Firms may decide to hold off on investing for a time, or banks might decide to reduce their lending for precautionary reasons. Those decisions reduce demand, reducing the amount of money flowing through the economy. Pessimism is a self-fulfilling prophecy, turning nervousness into reduced spending, into recession.

That is where policy-makers are meant to enter the picture. Central banks intervene in such circumstances to reduce interest rates. Lower interest rates are supposed to make saving less attractive; rather than put money in a savings account paying very little in interest, I might go ahead and buy that new dishwasher we need to replace the leaky old one: *voilà*, demand. At the same time, lower interest rates encourage households and firms to borrow – to take out a mortgage and buy a home, stimulating new construction, or to upgrade the office IT equipment. Central banks lower interest rates to boost demand and – perhaps more importantly – to overcome the self-fulfilling pessimism that created the shortfall in the first place.

Over the past generation, however, central banks have had to work harder and harder to keep economies on a healthy growth path. The process of closing the gaps that open up between what is earned and what is spent has become trickier and more fraught. Three overarching factors contribute to this new difficulty.

The first is the difficulty the world has had managing the era of rapid globalization. As we saw in the last chapter, emerging economies enjoyed rapid growth in the era of hyperglobalization by leaping into the global economy, joining up with global capital markets and supply chains. In the 1990s, emerging markets discovered that this new, highly globalized and highly financialized world economy could easily touch off crises in unprepared developing economies. Mexico, Russia and the 'tiger' economies of South-East Asia all faced crises at the hands of fickle foreign investors, who rushed into fast-growing economies to make a quick buck, then rushed out when the mood turned, leaving tumbling asset prices and bad debts behind.

Emerging economies learned that piles of foreign-exchange reserves – government savings held in the form of foreign currency or foreign-currency-denominated assets – could be used to provide a cushion and reduce the pain when jittery foreign investors became nervous and began pulling money out. A central bank that bought up loads of US Treasury stocks during good times could sell those stocks in bad times, to prevent its currency from crashing in value or in order to give dollars to struggling firms with unaffordable dollar-denominated loans. In the 2000s, reserve accumulation by crisis-fearing emerging-market central banks contributed to massive growth in global savings. Governments across the emerging world, and China first and foremost, bought up safe foreign assets – such as American Treasury bonds – in large numbers. These purchases were designed both to slow the appreciation of their currencies against the dollar (a weaker currency boosts exports, other things being equal), and to accumulate a defensive stock of safe assets. This effectively squeezed the consumption of Chinese consumers in order to build up a pile of reserves that could be tapped during times of global financial trouble.

The upshot of this reserve accumulation was growth in what Ben Bernanke,[2] during his time at the Federal Reserve, called a global savings glut.[3] Excess saving meant a shortfall in global consumption, in global demand. To prevent demand from tumbling, central banks needed to take what action they could to push markets to recycle those savings into new spending. Interest rates around the world sank to historically low levels as central banks struggled to cope.

But a second factor frustrated their efforts. These savings accumulated at a time when opportunities for profitable investment were often limited. Massive IT investment in the late 1990s supported strong economic growth, but such investment tailed off in the 2000s as firms puzzled over how to use their IT productively. The rising premium on know-how boosted the value of the skilled cities where firms and individuals were experimenting with more powerful computing and new communications technologies. Yet as we saw in Chapter 7, the increased value generated by productive cities did not translate into massive investment in new construction, due to the limits imposed by building-supply restrictions. Nor did governments take full advantage of low interest rates to invest in new infrastructure. Investment in new transport lines and networks might have helped growing cities like New York and Boston accommodate additional construction, thereby creating two productive outlets for accumulated savings. Alas, that was not to be.

Rising inequality – the third factor – exacerbated these difficulties. Households in the middle and lower portions of the income spectrum have what economists call a high propensity to spend. That sounds like an insult. It isn't. It simply means that because they earn less they have to spend more of their income to cover necessities. If you give a poor American household an extra $100, its members will probably buy more food, or replace worn out clothes or furniture: they will spend it, in other words, more or less immediately. Give a rich household an additional $100 and (assuming its members even notice) the money is far more likely to sit idly in savings. The rich household will not be rushing out to get that new television it has not quite been able to afford.

The share of national income earned by the richest households soared over the last generation. Wealth concentrated in fewer hands. Both trends pushed purchasing power into the grips of households with low propensity to use an additional dollar for consumption. In other words, more income flowed into the hands of people inclined to save a large share of their income. Given soaring savings and obstacles to productive investment, central banks therefore had to work harder to keep demand high enough to keep everyone employed.

DEBT AS A QUICK FIX

Over the last generation, markets bumbled their way to a solution to the problem of chronically weak demand. The answer was for the rich – those with lots of money but little interest in spending all of it – to find ways to lend money to those without much. From the 1980s on, rich economies devised cleverer and cleverer strategies for getting loans into the hands of households keen to borrow. In the early 1980s, total household debt in America came to less than 50 per cent of GDP. It then commenced a steady rise, to just under 70 per cent of GDP by the end of the 1990s. In the 2000s it skyrocketed, to close to 100 per cent of GDP on the eve of the financial crisis.[4]

Financial engineering facilitated the shift of money from those who had it to those keen to spend it. Banks came up with clever ways to package dodgy loans into securities that looked reasonably safe, but which promised a healthy return. The world's big savers, from China to the very rich, gobbled them up.

Yet governments also encouraged the transfer of purchasing power through massive lending. Low interest rates – the necessary consequence of attempts to keep demand on track when savings outstrip investment – reduced the cost of borrowing. Perhaps more importantly, regulators allowed risky financial practices to flourish, and even made regulation less restrictive in some cases as housing prices soared.

As house prices rose, everyone, from the bankers building lending models to the Joneses looking to buy their first home, increasingly expected them to keep rising. Rising values enabled much more lending. When prices are expected to increase, banks worry less about the prospect of default (since homeowners who get into financial trouble can easily sell at prices that cover the outstanding value of their mortgage). Homeowners with positive equity in their homes also took to borrowing against that equity to finance home improvements, or even consumption. From 2003 to 2008, for instance, outstanding home-equity loan debt in America more than tripled, from $200 billion to nearly $700 billion.[5] Remarkably, soaring home prices in the 2000s did very little for the net worth of most households, since they occurred alongside rising debt burdens.[6] Those debt burdens could not be sustained indefinitely.

The dramatic economic changes of the past few decades concentrated purchasing power in the hands of those – governments and households – with a low propensity to consume their lucre, at the same time that avenues to healthy investment were blocked up. Chronically weak demand therefore resulted. Central banks cut interest rates and eased financial regulation in an effort to solve the problem of weak demand by engineering a transfer of purchasing power, accomplished through massive lending by the rich to the rest. This lending put purchasing power in the hands of those who were eager to spend. The money began to circulate, and the economic slump eased – until the cycle of asset-price rises and borrowing came to an end.

When the music stopped playing, the world stood on the brink of the worst economic calamity since the 1930s. And the crisis introduced a new vulnerability into the system: as central banks worked to buoy demand, they slashed their interest rates to zero or, in some cases, to negative rates. Central banks are not entirely without options once rates fall so low. They can keep cutting, a bit, or they can print money to buy assets such as government bonds (a stimulative procedure known as quantitative easing). But these options are limited in a number of ways: as interest rates become increasingly negative, for instance, households have an incentive to shift more of their savings to cash – to keep their money in shoeboxes or safe-deposit boxes, where negative rates do not apply. Central banks themselves are also wary of acting aggressively in using these 'unconventional' policy tools: they worry about risks known and unknown.

When rates fall to near zero, as a result, policy tends to become more tentative and less stimulative than it should be. The secular stagnation trap becomes even more difficult to escape.

CRISES TO COME

A massive forest fire reduces the odds of another fire disaster in two ways. It leads, in some cases, to changes in forest management. More importantly, it clears away years of accumulated fuel. The global financial crisis sparked waves of financial reform, some of which has surely made the global financial system a little safer. It also wiped away (albeit in an extraordinarily painful fashion) the crisis fuel that had been accumulating: big household debts supported by high asset prices. The silver lining to America's nasty crisis was a return to relatively low levels of household debt.

The underlying imbalance between income and spending has not gone away, however. Indeed, it might be worsening. That is one significant reason why interest rates have been so low for so long, and are expected to stay at historically low levels for years to come.

The dynamics of the imbalance have changed, a bit. Big emerging markets are saving less. Many have, in recent years, been forced to use reserves to fight capital outflows. But rich-world firms have picked up some of the slack. In the years since the crisis, the profit rate among successful companies, in America especially, has stayed unusually high for an unusually long period of time. High profits ought to encourage lots of investment, as competitors attempt to muscle in on successful firms, and as successful firms draw down their war chests in responding to those competitive pressures. Instead, big, profitable companies have behaved a bit like emerging markets did in the 2000s: they have used good times to build up enormous cash piles.

Other structural forces are asserting themselves – the population is ageing, for one. Across the rich world, older societies are beginning to draw down their savings as the share of the population in retirement rises. But ageing also reduces the outlook for future growth and limits the need for new investment. So far, at least, the latter effect appears to be dominant: diminishing appetite for investment has more of an effect on the imbalance between saving and investment than the use of savings to pay for retirement.

The dematerialization of economic activity also reduces investment. The most productive activities around rely heavily on know-how – on social capital and software – rather than on great factories full of physical capital equipment. Processing power that grows cheaper by the minute can be harnessed to improve production across a growing share of the economy. The cost of cloud computing services continues to tumble, for instance. In the 1990s, at least, macroeconomists could count on start-ups to invest their money in big, energy-sucking servers; now they can rent what they need from Amazon or Google at a fraction of the cost. Savings pile up; potential uses do not.

Software is eating everything, and the creation of new software requires investments of time and social capital rather than mountains of money for plants and equipment. Capital *is* required for office space in social-capital-rich cities, however. But because of the continued difficulty in building new office space in such places, rising demand for offices in productive cities mostly pushes up real estate costs: London rent payments by technology firms become additional capital income and capital gains for the very rich, who spend too little of their marginal earnings to keep demand growing rapidly.

Indeed, inequality grew worse as a result of the Great Recession and its aftermath. The fortunes of the very rich took a beating in

2008–9 but recovered far more quickly than labour markets. Labour abundance has kept wage growth at very low rates, even in countries where unemployment rates have fallen back to normal levels. Workers' inability to increase their share of all income earned represents a continued drag on demand. Purchasing power simply isn't flowing to those with the greatest interest in spending. Demand remains too weak to absorb all available economic capacity. Central banks struggle to break rich economies out of the trap.

To escape the world of chronically weak demand, and of recurring crises, purchasing power has to find its way into the hands of those keen to spend and invest. There are a few ways in which this transfer might occur.

Governments could pick up the slack through direct spending. Increased transfers to cash-strapped households would be the simplest way to channel money to spenders, but massive programmes of public investment would also do the job. Government spending on highway repairs or construction of new railway lines or airports steers money towards construction workers and equipment manufacturers and steel-makers, and so on.

The government could finance this through taxation. Progressive tax systems could be used to direct money from households with low propensity to spend towards those likely to put the money to immediate use. But most governments could also soak up excess saving through increased borrowing. Savers around the world are happy to hold safe government debt at low or negative yields: that is how disinclined they are to spend. Were the government to borrow from those savers and use the proceeds to fund spending, that would help improve the circulation of money in the economy.

Alternatively, governments could finance transfers by printing money (or by issuing bonds that are subsequently purchased by the central bank). One might worry that monetization of current spending would lead to hyperinflation, of the sort experienced by Germany in the early 1920s, yet this risk is overstated. Hyperinflation typically results when governments are *unable* to raise tax revenue to pay for spending and are *forced* to turn to the printing presses. A government that is instead *choosing* to print money in order to stimulate a depressed economy can simply choose to stop printing once the economy is out of the trap. (Though in some cases even quite high levels of debt monetization prove too modest to kick an economy out of its rut. The Bank of Japan has been printing money like mad to buy government bonds; it now owns more than a third of outstanding Japanese government debt – a sum worth abut 300 trillion yen, nearly $3 trillion, or two-thirds of the entire annual output of the Japanese economy – and is on course to buy up most of the rest over the next decade or two. A decade ago, most economists would have sworn up and down that printing money on such a scale would lead quickly and inevitably to hyperinflation. Yet, in Japan, inflation remains just a bit above zero.)* A more permanent solution would be one in which workers are able to capture for themselves a greater share of the gains from economic growth. Were rich, productive cities to suddenly begin building massive amounts of new housing, that would help solve the demand problem in a few ways. Construction itself would boost demand, of course; resources would flow to carpenters, plumbers, welders and others with high propensity to spend. More importantly, it would slow or reverse growth in housing costs. That, in turn, would redirect the gains of growth that now flow to rich property owners into the hands of households with lower incomes. Workers would become less abundant relative to productive land, and could therefore capture more of the gains from growth.

Or to put things slightly differently: if workers could raise their bargaining power relative to other groups within the economy, they could grab a larger share of the gains from growth. That, in turn, would lead to more spending and more demand. It is not a

---

* Monetization and transfers might not be necessary in a world in which central banks were willing to tolerate higher inflation. If prices rose while wages did not that would be a hardship for workers, whose 'real' pay would fall. As real pay fell, however, firms would be more inclined to hire. Indeed, they could be so inclined to hire that they decided not to use available technology to replace workers, or might even scrap some of the labour-saving technology they had previously begun to use. As firms relied more heavily on labour, the share of national income going to workers as a whole ought to rise, while the share going to owners of capital might fall. More money in workers' hands should lead to more spending. Something like this seems to have happened in Britain, where falling real wages coincided with a rise in the labour share of income as employment leapt to all-time highs. The trouble, of course, is that most central banks consider themselves unable or are unwilling to push inflation up.

coincidence that central banks were faced with quite different challenges in the 1960s and 1970s than they were from the 1980s on. In the earlier period, workers enjoyed more bargaining power relative to owners: union membership was higher, and binding, generous wage deals with big firms were more common. As a result, rich economies found themselves grappling with high demand and high inflation rather than chronically weak demand and low inflation. High inflation comes with its own costs, of course, but it is much easier to manage, macroeconomically speaking. Global depressions do not tend to occur in high-inflation environments. In periods in which there is consistently enough spending to keep prices rising at (for example) 4 per cent per year, that spending acts as a sort of economic lubricant – contracts agreed in the past, over rents or loans, become more affordable over time rather than less, because the dollar value of prices and incomes rises relative to the dollar value of past obligations. In deflationary environments, on the other hand, such burdens loom ever larger, contributing to a cycle of cutbacks, defaults and further price declines.

A WIDENING GYRE

While the underlying conditions leading to chronically weak demand remain in place, the struggle to escape will grow harder, not easier, over time. Secular stagnation has a way of drawing additional countries into the trap, increasing the share of the global economy facing stagnant conditions and raising the gravitational pull towards the secular-stagnation black hole.

The world is stuck. Too many big economies are struggling to generate enough demand to use up all their available economic capacity. Economies stuck in such a position can achieve fast growth, however, by capturing demand from abroad: by boosting their net exports (exports less imports) to other countries. Countries can do this by squeezing wages and prices in their domestic economies, or by depreciating their currencies, both of which make the cost of goods and services cheaper in foreign economies.

These actions, if successful, place a demand drag on the rest of the world; some of the purchasing power in a country like America, say, is diverted to goods and services produced elsewhere – in Germany, for instance. In ordinary times, this would not matter much: America's central bank could respond to the diversion of spending to foreign economies by reducing interest rates to boost domestic demand. But in secular-stagnation economies, and especially those in which interest rates have fallen to near zero, the central bank cannot easily offset that drag. The more of the world's economies that find themselves in low-rate, secular-stagnation conditions, the more pressure is placed on those economies not yet stuck in the trap, since they are the remaining, reliable sources of demand. But as ever more of the spending power in those reliable sources of demand is sucked in by those stuck in the trap, the closer healthy economies are brought to the trap themselves. Widening trade deficits sap demand, and as central banks cut interest rates to offset that spending drain, their economies inch towards the edge of the zero-rate quagmire.

There are two ways to stop the trap spreading globally. Help could come when those economies operating at full capacity accept *excess* demand, in order to drag the world economy back to health: to spend so much that domestic firms running at full tilt cannot meet demand and imports therefore grow rapidly. In that case, the overheating economy would face significant inflation pressure and would run an enormous current-account deficit for a prolonged period of time. In the past, America might have been able to play such a role; it was big enough relative to the world economy and consumption-oriented enough to gobble up massive amounts of domestic and foreign production. But America's economy has shrunk in size relative to the world economy as a whole over the past few decades and it has itself descended into the secular-stagnation trap. Even if America were capable of single-handedly driving global recovery, it seems unlikely that the Federal Reserve would allow it to do so. Its mandate is to keep American demand from overheating; if wages and prices began growing at even a moderate pace the Fed would raise rates. The Fed's domestic focus prevents it from allowing America to generate the excess demand the world needs.

The other route out would be global coordination to generate sufficient global demand: joint central-bank easing and government spending around the world, designed to kick the world out of secular stagnation. The governments lucky enough to enjoy near-zero borrowing costs could invest heavily in public-spending projects, from infrastructure to bold research initiatives, in order to run big budget deficits; central banks, then, could print money to buy up the newly issued bonds. Sadly, global institutions seem not to be up to the task. Though some coordination was managed in 2008–9, when the global economy was on the brink of financial collapse, cooperation soon eroded.

The most direct historical parallel, the 1930s, is not an encouraging one. The world kicked itself out of the secular-stagnation trap back then through unilateral devaluations (as countries left the gold standard one by one) and through the stimulative power of massive military spending.

Weak global demand is an incredibly nasty and destabilizing force. It turns the global economy into a zero-sum battle, in which faster growth in one country often comes at the expense of another. It nurtures financial instability and crises, which also tend to fuel radical political movements. The longer the world remains stuck, the more likely things are to break apart in dangerous ways.

THE END OF ORTHODOXY

The tools of macroeconomic management on which most mature economies have come to rely were designed for a different kind of economy. Recessions in modern industrial economies generally occurred due to a particular kind of imbalance: in which firms and households in an economy all attempted to save too much at the same time. The imbalance was a temporal one. When too many people tried to shift purchasing power from the present to the future, the economy slipped into recession. Governments, and central banks, then needed to step in to smooth spending out: to encourage people to bring some of that deferred spending back into the present.

Today's imbalance is different. It stems from the fact that purchasing power is concentrating in the hands of particular organizations and individuals rather than in different time periods. The different character of the world's macroeconomic problem has become increasingly obvious over the last generation, yet governments have clung tight to old orthodoxies in their efforts to keep the global economy on track. But the old orthodoxies are not going to fix the problem. The rules are not what the people in charge think they are. Big deficits and debts don't lead inexorably to soaring interest rates. Printing money to buy up government debt doesn't lead inexorably to hyperinflation. A central bank that targets low and stable inflation rates does not automatically keep the economy on a healthy growth path.

Orthodoxy needs to shift to accommodate two underlying truths of the digital, globalized age. First, any economy that is linked into the global financial system cannot escape the influence of global demand – and the global balance of saving and spending. And second, keeping demand growth on track requires the redistribution of purchasing power from savers to spenders. That redistribution will be achieved one way or another: through competitive depreciations, or through inflation, or through soaring debt, or through direct government transfers.

Direct transfers are the least dangerous way to fix the problem. But aggressive redistribution of resources – however it occurs – is a radical policy. Governments do not adopt radical policies until forced to do so, by crisis, and by the sweeping change in political priorities that eventually occurs in response to economic revolutions.

It will take crises of greater severity than that of 2008–9 to generate such political change. Unless the world is very fortunate, there are more economic maelstroms ahead.

4

From Abundance to Prosperity

IO

Why Higher Wages are so Economically Elusive

In the *Star Trek* universe, the scarcity problem has been solved. The manipulation of atoms has been perfected, to the point that anything anyone wants can be 'replicated' for them at a moment's notice. Freed from material shackles, the citizens of the Federation are able to elevate themselves, to enjoy the finer and nobler things in life, and indeed to boldly go where no man or woman has gone before.[1]

Humanity might eventually arrive at such a place, where abundance is nearly endless and only the heat death of the universe threatens our good-natured fun. Sadly, that is unlikely to be a concern for those of us alive now. As rapid as the pace of technological progress has been, we are still creatures bound firmly by scarcity. In our world, technology will often be freeing, but the allocation of resources will continue to matter, and in many cases technological progress will only be as liberating as the organization of society allows it to be.

For now, the world economy operates on a framework very much rooted in an industrial, scarcity-bound world. The interaction of that world with the technological advances of the digital era have landed labour in a trap. The digital revolution generates fantastic labour abundance; that abundance contributes directly to downward pressure on the wages of the typical worker. It also reduces the bargaining power of labour relative to other, scarcer factors, allowing those factors to capture outsize shares of the gains from growth.

We might not care so much about these inequities if the digital revolution were reducing the costs of all the many things the typical household wants to buy, from steak dinners to adequate housing to a top-flight university education. But cost reductions have so far been highly uneven: massive for some things, such as digital entertainment; completely absent for others, such as homes in nice neighbourhoods.

And so stagnant wages end up mattering an awful lot. Low pay for the great mass of workers is distributionally unfair. It undermines support for the market-based economic system that enables sustained economic growth. Low pay also reduces the incentive to invest in technologies that boost the productivities of less-skilled workers, or which substitute for less-skilled workers. In a very narrow sense, that is a blessing: it tempers labour abundance and ensures that there continues to be employment for millions of workers. In a broader sense, it is an enormous problem; continued productivity growth is ultimately the route to lives of greater comfort for all of humanity: it is how humanity contrives to produce more from less, so that all can have more.

To achieve a better, more just society, incomes must rise. Not just the incomes of China's middle class and the rich world's 1 per cent, but those of modestly skilled workers the world over. The share of income globally that flows to labour, and the share of labour income that flows to those in the bottom 90 per cent of earners, ought to level off and rise back in the direction of the shares earned a generation ago: when the top 1 per cent captured less than 10 per cent of national income rather than a share between 10 per cent and 20 per cent, as they do now in America and Britain.

But achieving higher incomes is a fraught business, both economically and politically. This chapter will consider the economics of the problem; the next will take on the politics.

THE DIFFICULTY IN RAISING PAY

The most straightforward way to lift incomes is simply to raise them directly. Governments often set a minimum wage; that minimum could be raised considerably higher. Many governments also subsidize wages (such as through the Earned Income Tax Credit, in America, or the working tax credit, in Britain); those wage subsidies could be raised. In many countries, the maximum subsidy for working adults is not especially generous; in America, a household without children receives about $500 at most. Governments could go beyond that: to set a minimum income level, for instance, so that all household incomes below that level (including those which are zero, pre-subsidy) would be topped up to the minimum. Or governments could pay a basic income to *all* citizens. A few countries – including Finland and the Netherlands – are experimenting with more generous wage subsidy programmes that would approximate a basic income.

The main difference between a minimum wage and a policy of wage subsidization is in who pays for higher wages. In the case of the minimum wage it is the employers (and, to the extent that the employers enjoy market power and can therefore tack the cost of higher wages on to prices, consumers). For wage subsidies, the cost is mostly borne by taxpayers.[2]

Both approaches present trade-offs: difficult ones for society to manage. Take the minimum wage rise first. Minimum-wage increases unambiguously boost pay for those people who previously worked at wages below the new minimum and who remain in employment after the rise. Interestingly, evidence also suggests that minimum wage increases can push up wages for workers earning *above* the minimum, perhaps because the measures firms take to manage higher labour costs – investing in monitoring to improve effort, for instance, or increasing training – affect workers other than just those labouring at the minimum. Yet there is inevitably a down side.

Studies of modest minimum wage rises turn up mixed evidence of their effect on employment: in some cases firms reduce their hiring, while in others profits shrink a bit. But modest minimum wage rises aren't really what we're discussing here. To achieve meaningful increases in the labour incomes of those at the low end of the earnings spectrum, minimum wages would need to rise considerably. Yet this book argues that employment has only continued to grow, to absorb the rising number of participants in the global workforce, because wages have fallen or stagnated for many workers. Low pay allowed firms to hire such workers to do low productivity tasks.

If that is correct, then much higher minimum wages would necessarily lead to large reductions in hiring and employment. Consider the American economy for a moment, where a movement to raise the national minimum wage to $15 an hour is gaining momentum. The median hourly wage in America, in 2014, was about $17; half of workers earn more than that and half earn less. Just under half of workers – about 66 million – work in occupational categories that pay a median wage of less than $16. Over 36 million people, more than a quarter, work in occupations paying a median wage less than $15. In the food preparation and services category, which accounts for more than 12 million American jobs, 90 per cent of workers earn $15.12 or less. Indeed, half, or just over 6 million workers, earn $9 or less.[3]

Were the minimum wage to rise to $15, some share of those now earning less than that amount would enjoy a raise, paid for out of current profits or through higher prices. Some workers would work harder to justify the increased pay; firms would invest in training and reorganization to boost the productivity of others. Yet many other workers, probably numbering in the millions, would face job loss as firms adjusted their business models to make themselves less reliant on cheap labour. Fast-food restaurants would become less common and more expensive. Those that stayed in business would find ways to reduce the labour-intensity of the restaurants.

Is there no way around this? Might not the workers earning higher minimum wages buy more things with their larger salaries, raising demand for other goods and therefore for labour? Possibly – if the higher minimum wage were adopted as part of a plan, coordinated with the central bank, to raise demand. Yet such a plan would only work because everyone in the economy would expect higher spending to translate into higher prices, ensuring that real, or inflation-adjusted, wages did not rise by very much.

In a world in which low pay is the main mechanism through which less-skilled workers are kept in employment, mandated higher pay necessarily leads to lower employment. Meanwhile, efforts to boost productivity to adapt to substantially higher minimum wages simply exacerbate the problem of labour abundance: as firms work to wring more production from fewer, more expensive workers, they add to the world's labour glut, placing downward pressure on wages broadly. And higher minimum wages do nothing to boost the incomes of those unable to find work.

Minimum or basic incomes are in some ways more promising. The benefits are clear enough: those eligible for the basic income will earn the basic income and no less. Depending on how they are implemented, basic incomes could be simpler to administer than other welfare programmes. They could be used to encourage pro-social behaviour in those unable to find work; governments could require those earning a basic income to either work or to provide public service of some kind in order to earn the minimum. Freed of the need to generate a living income, creative types could use the income to support a socially valuable (if frequently unprofitable) life producing art or music, or craft goods and services. Entrepreneurs keen to open cafes or start consulting businesses that might not, initially or ever, pay enough to provide a living wage after business expenses would be freed by a basic income to take the plunge.

Importantly, because governments, rather than firms, pay for the wage top-up, firms are not given the incentive to economize on labour as a result of the policy. The higher taxes needed to pay for a basic income could conceivably squeeze firms in other ways, but most rich economies have scope to boost the efficiency of their tax systems. The trouble occurs instead on the worker's side of things.

People generally work because they need money to pay for necessities. Yet those without work are not, in most cases, left completely destitute. Support from family, charities and the government provides the unemployed with the means to stay alive without a job (albeit in straitened circumstances). The poorest workers in an economy face a choice between finding a job and living off that meagre alternative income. As the wages available in the labour market fall, the meagre alternative looks increasingly attractive. In the same way, as the alternative income becomes more generous, the more appealing it is to opt out of difficult, low-pay work.

When a government chooses a basic income level, it must be aware of the trade-off: the more generous the income, the more workers at the low end of the income spectrum will choose to forego work entirely, because the pitiful wages available for market work aren't worth the time and effort it takes to earn them. On the other hand, very low basic income levels ensure that less-skilled workers are kept in penury, earning wages that do not keep up with average income growth. A basic income, paid to all citizens regardless of work status, would be more expensive than a minimum but would slightly reduce the incentive to drop out of the labour force entirely, since finding a job at a low wage would not reduce the basic-income benefit. Even so, for a sufficiently generous basic income, many people would opt not to engage in work.

Many proposals for basic income programmes attempt to have the best of both worlds, by requiring work of some sort to qualify for the minimum.[4] Such requirements might be the only way such policies gain the political support to become law, but they suffer from significant weaknesses. They are sure to be costly to enforce, and could lead to millions of people wasting time doing pointless work simply to earn the money they need to survive. They are also illiberal; in a world in which technological abundance makes the labour of a large share of less-skilled workers essentially unnecessary, it seems churlish to require those already stuck earning the lowest incomes in society to jump through hoops for those incomes. Perhaps most importantly, they are morally untenable; if large numbers of workers rejected the requirement to work at menial tasks, society could not commit to letting them starve to death. (Nor should it!)

One might argue that the disincentive to work provided to labour by a basic income is part of the point; it clears essentially unnecessary labour out of the labour market, freeing them to enjoy their lives, raising pay for more productive workers, and encouraging firms to keep investing in labour-saving technology. It is possible that, over time, as technology continues to improve and economies grow, and assuming the basic income rises with average GDP per person, that ever more of the workforce would find it attractive to abandon work-by-necessity for other pursuits. The basic income could – could – be the means by which humanity's leisure-filled technological utopia is eventually, gradually realized. If the politics allow it, that is. At least initially, such a plan would create a class of idle workers, made up mostly of those with the lowest productivity levels, supported by the productive rich.

A basic minimum income forces governments to make difficult choices: to allow a large share of the workforce to avoid participation in the labour market, or to keep a large share of the workforce in penury, or to spend heavily to make sure a large share of the work-force completes an adequate amount of time-occupying work.

Both policies are and will remain a part of the policy-maker's toolkit. Both will be implemented reasonably well in some cases, and quite badly in others. Both will be used in better or worse fashion according to the political dynamics that will be discussed in the next chapter.

WORKING SMARTER

An alternative to arranging for governments or firms to give workers more money is to make workers more productive, and therefore to enable them to demand more money for themselves. Increased educational attainment helped to tame the industrial economy on behalf of workers, by enabling an ever larger share of the workforce to find ever more sophisticated technical or analytical work, reducing the glut of less-skilled labour competing for menial jobs. Policy-makers are inevitably enthusiastic about education as a solution to whatever happens to be ailing the economy (as opposed, say, to dramatic reform of the welfare state).

More education, especially in emerging markets, would be a good thing for lots of reasons. Individuals benefit from education not just because it improves the set of economic opportunities they face, but also because it helps them make better personal financial decisions, or improves the odds of meeting and marrying a well-educated spouse. So one should not necessarily conclude that humanity is educated enough simply because more education would probably not solve the problems described in this book.

Where increased education alleviates fundamental growth bottlenecks – by increasing the numbers of knowledge-frontier-expanding engineers and scientists, for instance – it can increase growth and the size of the economic pie to be distributed. If other economic factors are the bottleneck, however, then education mostly boosts the fortunes of some groups by reducing the relative scarcity of others: nurses who train to become doctors earn more, but they also reduce the bargaining power of the existing pool of doctors by increasing their numbers. And if increased education raises effective available labour by enough – if it mostly adds to the abundance of effective labour available in the world economy – then it might simply reduce the bargaining power of labour as a whole relative to other factors in the economy, such as land or social capital. In the absence of other policies, in other words, it could potentially, and quite counterintuitively, leave workers as a whole worse off.

That argument is less applicable the less developed an economy is. Though rich economies are probably close to producing as many top engineers and researchers as they can from their domestic populations, there are vast numbers of clever men and women across emerging markets who lack access to quality education (and who are otherwise economically and socially constrained) and who therefore cannot contribute intellectually to society at their full potential. Those people represent scientific breakthroughs unmade, inventions not invented, world-changing companies not created.

Emerging-market societies could more obviously use more skilled populations; they need more trained professionals of all sorts: doctors and engineers, lawyers and financial professionals, and trained civil servants. Addressing the real skill scarcities that constrain development in poorer economies is no simple matter: quality education begins at a young age, which suggests that developing economies need school systems that are both enormous and of high quality – a tricky thing to manage in countries which almost definitionally lack the social-capital infrastructure needed to support complex institutions. International organizations and philanthropists are looking to fill the gap with technology. Educational technology will probably do an awful lot of good for well-disciplined young students who don't need much handholding to succeed, or for older students who have already received some basic education, but it probably cannot substitute for skilled primary-school teachers.

In general, no solution for boosting the educational levels of children in poorer developing economies is likely to be nearly as effective as immigration to rich countries. Yet this truth begins to illustrate why education cannot hope to solve the global problem of labour abundance. A shortage of skilled doctors is not a growth bottleneck in rich countries. Doubling the number of doctors working in America – either by increasing the educational attainment of native workers or by accepting immigrant workers – would not generate an appreciable increase in American economic growth. It would make doctors more abundant relative to the labour force as a whole: doctor bargaining power would fall, and doctor incomes would rise more slowly – or would perhaps even fall a bit. The doctor doubling would therefore be good for the *new* doctors (whose salaries would rise), for American consumers (who would enjoy a one-off rise in real incomes thanks to the drop in the cost of medical services) and for the managers and owners of healthcare firms (who would be able to reduce their labour bill and boost profits). Those benefits would derive, for the most part, from the reduction in bargaining power and pay of the original doctors.

This is not just true of the most academically demanding professions. Training more rich-world workers as electricians might well be a sensible thing to do, given the scarcity, in many cities, of trained technicians, yet the primary effect of this sort of training would be to reduce the scarcity of existing electricians relative to the customer base. New electricians would enjoy a wage rise, electrician wages as a whole would fall, and some of the gains of the shift would accrue to the households and firms looking to hire electricians, who suddenly benefit from increased choice at reduced cost.

A big boost in the educational attainment of less-skilled workers around the world would probably be good for less-skilled workers; data continue to show a sizable wage premium for workers with college degrees relative to those without. Yet gains captured by those moving up the educational ladder might come at the expense of those already on higher rungs. The premium earned by college graduates might shrink, even as college-graduate incomes stagnate or fall: indeed, since 2000 or so, college graduates in rich economies have not enjoyed meaningful wage rises. Depending on how productive the newly well-educated workers actually are, increased educational investment, by increasing the total effective labour in the world economy, could boost worldwide labour abundance, *worsening* the distributional problem for labour relative to capital and land.

That is not to say that efforts to improve educational attainment around the world are a bad idea. It *is* to say that education is almost certainly *not* a solution to the problems identified in this book.

CAPITAL GAINS

It might be possible, however, to make workers more productive while also reducing their relative scarcity, by dramatically increasing the amount of capital in the world economy. The world is awash with savings at the moment, which would seem to suggest that the problem has already been solved. Yet much of that capital is not being used productively because of the structural inadequacy of demand resulting from stagnation in labour incomes. Governments might address that inadequacy by facilitating massive investment in *social stuff*: in productive capital that benefits all of society.

The lowest hanging fruit in this category is that which might be harvested now, were it not for exclusionary rules. Much more housing and office space could be built in highly productive cities, if zoning and other regulations were eased to allow for it. To alleviate the potential congestion that might result, societies could invest massively in infrastructure: in new road, rail and air transport, in electrical grids and water systems, in capacious new broadband connections, and in grand new public parks and other social goods. Many of the world's richest cities are choked by inadequate infrastructure. Given rock-bottom interest rates, it seems absurd that places like New York, London and San Francisco aren't receiving extraordinary new investments in that infrastructure: in new airports connected by fast rail with central cities; in vast new rapid transit networks creating new nodes around which higher density housing could be built; in new water, sewage, electrical and communications networks; and so on.

The capital investments with the greatest potential productivity are those in poor economies, where the level of capital per worker is lowest. Raising capital per worker in such places is hard, however. Developing-economy financial markets are often poorly developed and easily overwhelmed by inflows of foreign capital. Where money can flow in, it might be subject to expropriation by corrupt governments, or inadequate maintenance, or other problems associated with the weak social capital infrastructure. As with education, the most effective way to raise the level of capital (physical and social) per worker globally is through increased immigration to rich economies.

THE ELUSIVE POTENTIAL OF IMMIGRATION

Rich-economy social capital is scarce, globally speaking: most countries lack it, which is why most countries are not rich. Financial and infrastructure capital is scarce relative to labour, in rich and poor economies alike. But *savings*, globally, are abundant: investors in both rich and poor economies are keen to stash their money in safe assets. Ideally, all of these things could be brought together: savings could be invested within economies with strong institutions, while also funding investments with the highest yields, and satisfying markets' appetites for safe assets.

The rich world could shepherd the global economy towards that ideal outcome by allowing lots more immigration from poorer places. Immigration would enable richer places to 'export' their strong social capital to poorer places (by bringing large shares of the populations of poorer places into countries where strong social capital dominates). That would naturally deepen social and financial capital per human worker. Global savings could also be mobilized to invest in additional infrastructure in the rich countries accepting new workers. The appetite for safe, rich-country government debt is nearly insatiable.

As with education, however, this sort of solution – though it could easily improve millions of lives – would not necessarily reduce labour's abundance relative to other factors, even with truly massive amounts of investment in infrastructure, housing and equipment, and the extension of rich-world social capital to immigrants. But the relocation of tens or hundreds of millions of capable workers to countries with much stronger economic institutions than those they left behind would generate an enormous humanity-wide increase in labour productivity: that is, a large rise in the effective labour available to firms.

A flood of new workers would be especially likely to reinforce labour abundance if *firm* social capital remained a bottleneck. That is, if the most successful companies enjoyed comfortable market positions, unthreatened by competitors, as a result of their highly evolved, highly effective social capital structures, *and* if those successful companies were uninterested in boosting activity and employment simply because more labour was available, then the flood of new workers into rich economies could lead to much more wealth concentration in the hands of those sitting atop those successful firms.

Even so, expanded immigration to rich countries could make the world as a whole a much richer, and more equal, place. While workers already in the rich world would probably experience continued slow wage growth as a result of immigration, the migrants themselves should enjoy a substantial rise in income. Relocating a larger share of the world economy inside countries with strong political and economic institutions might generate other benefits as well, from reduced transaction costs to a reduced ability for large firms with market power to play economies off against each other in search of the most lenient possible tax and regulatory treatment.

Yet the distributional concerns of those already living within rich countries would hardly be alleviated by this sort of plan. And the decision to allow or not allow such migrant flows rests in their hands.

PROXIMITY VERSUS EXCLUSION

As the above examples ought to make clear, there is no trade-off-free fix to the labour abundance generated by the digital revolution and the social challenges that result. Yet there is one overarching trade-off affecting the economic and political choices made by those in privileged positions in the world economy: the trade-off between proximity and exclusion.

Within rich economies, one need not work at an especially high productivity level to earn a good salary. It is enough to work in *close proximity* to those who work at high productivity levels. Lawyers and barbers alike in the San Francisco Bay Area earn more than their professional peers in Appalachia or Albania.

Imagine, for a moment, one of the more idyllic conceptions of life in a world of technological abundance: one that is already emerging in many of the world's richer cities. In a surprisingly large number of service jobs, low productivity is not a negative; it is the most marketable aspect of the work. Someone looking to buy a piece of art will not lament high prices and wish technology would come up with better ways to produce painted pictures at less expense. The time spent by the artist, and indeed the cost of the piece, is part of the attraction. The same is true of a maker of artisanal cheese, who sells their products at a steep mark-up in the local farmers' market. The food industry has become very good at producing massive amounts of cheese – even of a relatively good quality – at minimal expense; the attraction of artisanal cheese is quality, yes, but also the very fact that it was *not* produced industrially.

The trendiness of artisanal goods and services among well-heeled, cosmopolitan sorts is easy to lampoon. Everyone enjoys a chuckle at the absurdity of the Brooklyn hipster in *The New York Times* profile munching on artisanal jellybeans.[5] Yet there is something appealing about the role artisanal production plays in a polarizing economy. For producers the work can be intensely satisfying: hands on, with a clear end product of generally high quality. What is more, artisanality cleverly manages to shift income from the rich to the rest.

The North Carolina where I grew up, for example, was highly agricultural outside the few larger cities. The southeastern portion of the state was blanketed in fields of peanuts and tobacco, among other things, and massive hog pens. The farming was industrial in nature, store prices were low, and while hog magnates could earn a good living, agricultural life was generally one of hard work and very low incomes.

But the economies in the large cities were changing rapidly. Raleigh, my hometown, is part of a thriving tech hub that has enjoyed explosive population growth over the last generation. As the city has grown its population has become better educated, richer and more cosmopolitan. Trendy bars and restaurants now populate the once-barren downtowns of the hubs of the 'Research Triangle', of which Raleigh is one corner. And the sorts of ingredients demanded by Raleigh's new residents, and its growing ranks of bold and talented chefs, are quite different from what farmers in the state have traditionally produced. Slices of an industrially produced North Carolina ham might sell for a few dollars in a city grocery. A ham produced artisanally and sliced into high-quality prosciutto might fetch ten times as much or more. As Raleigh residents grow richer, they become more interested in the back-story of the food put on their table. Chefs want to chalk on their blackboards that their tomatoes are locally sourced. The result is a slow and modest but nonetheless real reinvention of portions of the state's agricultural economy, which has allowed some producers to back away from industrial processes and, in doing so, to capture a share of the wealth being generated in the tech offices of the Research Triangle.

The artisanal market extends well beyond the world of food. The craft beer revolution in America and elsewhere is another example, while there are more and more firms emerging that sell fine clothing items that are expensive and trendy not because they are the work of a coveted designer, or because they are bespoke, but because they are artisanal: produced (often locally) in conditions advertised as dramatically different from the sweatshop environment of the garment factories in Bangladesh or Indonesia. There are producers of artisanal furniture, bicycles, jewellery, shoes, and so on. Maybe, in some wonderful future, we will all spend time lovingly producing craft goods for each other, while technology provides all the basic things we need at ultra low cost.

But maybe not. To profit from craft production, one must be proximate to wealth. A producer of craft goods in Mumbai who expands her customer base from the nearby slums to rich economies by taking advantage of online markets and logistics sites can raise her income enormously, yet she would do better still if she could move to a rich American city: even in wage-stagnant America, the labour-market alternatives available in the rich world ensure that artisanal producers in US cities are far better compensated than workers in emerging economies.

And the richer the city for which the craftsman produces, the higher the available income. Geographic proximity matters; someone who lives close enough to the Bay Area to tend Bay Area bars or coif Bay Area heads will have higher earning potential than a similar person who lives in a Californian city too distant from the Bay Area to have a feasible commute into that city. But social proximity is perhaps even more important than this. In many places around the world just a few miles of distance separates economies with massively different average income levels: Miami and Havana, for instance. National borders are critical lines of differentiation, given the importance of national social capital and the national institutions created to support it, but they are not the only ones that matter. There is a profound social gap between the neighbourhoods in the eastern and western halves of Washington, DC, for example.

What does social proximity get you? It provides access to critical formal and informal economic and political institutions: markets, legal regimes, entrepreneurial norms. It provides access to information networks, which assist in learning – about techniques, market conditions and all sorts of economically useful details – and facilitate the creation of good matches: between buyers and sellers, hirers and hirees. And it confers social affinity, which allows for full participation in formal and informal institutions of social support, 'permission' to participate in and benefit from society, and so on.

Yet, as we have seen, proximity is not the only route to riches; scarcity also matters. Craft society might quickly run into trouble due to sheer numbers: there may be too many would-be craftsmen to allow wages in the artisanal sector to stay elevated. The way to ensure scarcity for many craft practitioners is through exclusion. It is worth noting that crafts have nearly always relied on occupational protections, from guilds hundreds of years ago to licensing rules today, to limit entry and prop up wages. Occupational licensing is again on the rise. And that is just one of the ways in which workers seek to protect their wages by limiting competition.

Exclusion is all around us. It is most visible and dramatic when embodied in restrictions on migration across national borders, but that is hardly the limit of the use of exclusion to protect economic status. NIMBYism is a potent exclusionary force, which uses limits on development and high housing costs to shut outsiders out of neighbourhoods, school systems and dynamic economies.

Corporate power is also exclusionary. That may seem counter-intuitive – competitive markets ought to be a force for leanness in production – yet consolidation has been the rule in the American economy over the last few decades: across most industrial sectors the top firms enjoy higher market shares now than they did in the 1990s. Even trailblazing internet firms, heralds of the digital economy, are covetous of market power: gobbling up potential competitors, using litigation, regulation and bullying to secure exclusive access where they can. Facebook has bought up would-be social-network rivals, the better to build for itself something like a parallel internet. Amazon is ruthlessly acquisitive and has used its market power to push around publishers and sellers of all sorts. Uber has adopted driver rules that appear to be aimed at reducing operators' ability to work for rival firms.

Corporate power can arise through government favouritism or subsidy. It can come about as a result of natural monopoly: when the initial investment in a network confers on a firm a low cost basis against which new entrants cannot hope to compete. It can occur through mergers and acquisitions. But it pays off handsomely for the winners. In the 1990s, the most profitable firms earned returns on investment roughly three times those firms in the middle of the distribution; in recent years that figure has risen to *ten* times.[6]

What is true for firms applies just as much to labour, and the lesson is not lost on workers. Proximity and exclusion – two routes to greater incomes for labour – are diametrically opposed. More of one necessarily means less of the other.

Unhappily for humanity, politics pushes strongly in favour of exclusion over proximity as a means to income-protection.

11

The Politics of Labour Abundance

The last generation, during which the digital revolution's first power-ful effects made themselves felt, was an era of remarkable political moderation and consensus. The period began, in the 1970s and 1980s, with a liberalizing impulse across a broad range of countries, from Britain and America to China and India. While Thatcher and Reagan cut tax rates and squashed unions, Deng Xiaopeng trod cau-tiously towards limited tolerance of markets and foreign trade. The era of consensus continued with the collapse of communism in Russia and Eastern Europe, which prompted Francis Fukuyama to muse that 'the end of history' had arrived with the global ascendance of liberal democracy.[1] As global markets integrated, politics in most rich democracies coalesced around support for market-oriented econo-mies, global openness and progressive social goals. It was a pleasant sort of era for the cosmopolitan, technocratic elite: the believers in the notion that markets, lightly tended, offered the best route to global prosperity and peace.

This political era is at an end.

Around the world, dissatisfaction with the fruits of economic inte-gration fuels inward-looking political movements: protectionist in some places, separatist in others. Some politicians find themselves able to gain traction by playing identity politics or by criticizing insti-tutions of liberal democracy. Many succeed through withering critiques of the elites who minded the tiller over the last few decades. Faith in markets and their ability to generate broad-based growth has been shaken.

While the financial crisis and recession strengthened these political strains, they had begun to develop well before 2008. In America, for example, party polarization emerged gradually. Ideological sorting in the 1980s and 1990s laid the groundwork for the dynamics that followed, in which more radical elements within the parties increasingly set their agendas. In Europe, far-right and separatist parties became an increasingly persistent and occasionally disruptive force by the end of the millennium and the start of the next, as when Jean-Marie Le Pen, a French nationalist and anti-immigrant politician, advanced into the second round of the presidential elections in 2002, shocking Europe.[2]

Radical political movements have persisted for too long and enjoyed too much electoral success to be written off as a short-lived reaction to an economic downturn. America's increasingly virulent polarization thrives even as the American economy racks up impressive economic growth. Modest tweaks to welfare states or curbs to immigration might neutralize radical political movements, but they probably won't. Instead, these political disturbances reflect the opening exchanges in a long societal negotiation over just what the state and the economy ought to do, and for whom, in the digital era. If the industrial era is any guide, this negotiation will last for decades to come, and will occasionally result in dramatic, and possibly even violent, changes to the structure of global politics.

The outcome of this negotiation will depend in part on what the typical voter – or citizen, in less democratic countries – determines he or she wants out of life. But the evolution of political institutions will be just as important: it will determine how political priorities are expressed. The future is unpredictable, but we can sketch out some of the dynamics that will influence its progress.

YOURS, MINE AND OURS

Let's first talk about pies. Politicians like to say that it is better to find ways to make the economic pie larger than to argue over how to slice it. If the pie is the same size from year to year – if national income does not grow – then one person can only be made better off if another is made worse off. If a new person joins that particular society, then slices must be cut more thinly for all. If the pie is made larger from year to year, on the other hand, then there is more income available for everyone. An increase in the size of the pie – economic growth – creates at least the possibility of making everyone in society better off.

The implication of this metaphor is that the nature of economic growth shapes political priorities. The typical voter desires to improve their standard of living over time. That may yield support for growth-boosting policies, if voters can be persuaded that such measures are most likely to yield larger pie slices over time. Or it could yield support for redistribution. Political momentum for economic liberalization in the 1970s and 1980s emerged as typical voters lost confidence in the ability of more statist economic policies to raise long-term living standards.

The outcome of that liberalization differed substantially across countries. In China and India, liberalization delivered on its promise. In China, especially, a generation of rapid growth succeeded in elevating a large middle class out of poverty. China's economic pie grew massively. Distributional effects scarcely mattered, given the extraordinary growth across all slices.

In the rich world, things worked differently. In 2014, the inflation-adjusted income of the typical American household was just 7 per cent higher than it was in 1979. By contrast, the income of a household in the 95th percentile of the income distribution grew 45 per cent over that period.[3] During this era, the economic pie grew substantially: America's economy more than doubled in size. Yet because so much of the growth flowed to the richest households, distributional effects swamped growth effects in determining the change in living standards for the typical American. The typical voter very plausibly could have been made better off by a set of policies that reduced overall growth, but which steered much more of the benefits of growth to those outside the 1 per cent.

The way in which slices of the economic pie are cut determines whether the typical person perceives growth-boosting (or pie-increasing) policies as likely to yield that person a larger slice in future (or indeed, whether such policies yield a larger pie at all).

One could say something similar about attitudes towards the sharing of the pie with newcomers. With sufficiently rapid growth in the size of the pie, the fact that more pieces must be cut from it to accommodate growth in population (from immigration, say) is not especially important, because broader growth means that individual pie slices are nonetheless getting bigger from year to year. If growth in the pie slows, however, or if it is perceived to slow by typical voters as a result of the growth in the incomes of the very rich, then the additional slices being cut for newcomers suddenly grow in political salience. Whether or not rigorous analyses bear the argument out, the typical voter perceives that slices cut for newcomers are made at their expense.

All of which is to say, when median incomes are stagnant, distributional political arguments increase in salience. Arguments over distribution necessarily incorporate debate over which people can reasonably claim *any* share of the gains from growth. Any debate about national distribution necessarily incorporates theories about the economic rights of outsiders relative to insiders.

Inequality in the rich world has grown for several decades. Why have politicians been slow to make hay of this? Studies of redistribution across rich economies show there has been far less of it than one would expect from a world in which the fortunes of the typical voter shaped policy.[4]

This gap might be due, in part, to the fact that technology has provided some compensations to those with stagnant incomes, which have not shown up in the income data. A generation ago, the median household may have earned the same income, adjusted for inflation, that it does now, but it was not able to tap into a global information flow packed full with endless free entertainment. In some economies, and especially America, the real purchasing power of many households was given a boost as a result of migration from the expensive metropolitan areas, at the heart of the economy, to cheaper and more economically peripheral ones. In addition, inequality in *consumption* has not risen as much as inequality in incomes, thanks largely to a long period of growth in consumer indebtedness; households borrowed to prop up consumption – until 2008, that is, when households were suddenly forced to de-leverage in rapid fashion.

Yet the main reason that it has taken so long for politics to adjust is that political systems are stubbornly resistant to change. Political parties are massive social institutions that persist for decades or centuries by building and maintaining connections with collections of smaller interest groups and individuals. Those people and interest groups come to define their political identity through their association with the party. Identities are not rewritten on a whim.

As the economic fundamentals within a country change, particular interest groups – labour unions, for instance, or industry groups, or rich financiers – find that some of their traditional policy interests no longer line up with the priorities of the party they traditionally support. Yet that does not always immediately translate into political realignment. The economic interests of the disaffected groups may temporarily be overridden by other, non-economic policy concerns, such as wars or crime. Party leaders may attempt to placate dissatisfied groups with policy sops, to buy their continued allegiance. And it may simply take time for the leaders of the disaffected group to begin to think of their interest group as out of place within the party's internal coalition.

Even after that realization, political realignment still takes time. The legal scholar David Schleicher writes that, across the world's democracies, the last few decades have generated 'the rise of swaths of fundamentalist or expressivist opinion in parts of the electorate'.[5] This rise manifests differently across countries depending on the structure of the political system. In America, where the political structure strongly favours a two-party system, the rise of fundamentalist factions led to partisan polarization: to ideologically coherent parties rigidly opposed to each other (and backed by similarly polarized voting blocs, interest groups, donors, and so on).

Disaffected groups that might in other political systems have left the Republican Party, for instance, instead found themselves with no choice but to stay within it, and to wage an intense campaign to drag the ideological consensus within the party in their direction. In a different political system, the Tea Party faction might simply have broken off from the Republicans, competed in elections on its own terms, and joined occasional governing coalitions. In America, splitting off would have meant political irrelevancy, so the Tea Party waged an internal campaign – targeting Republicans with vaguely moderate sensibilities in primary campaigns, recruiting and funding champions – to seize control of Republican institutions.

Intense, intra-party battling could eventually lead to irreconcilable differences, prompting a major party shake-up (or break-up). Such shifts are extremely rare in American politics, however. It is more probable that ascendant ideological camps within the Republican Party, such as those with nativist passions, will come to dominate the party leadership, displacing the former establishment bosses. And the former establishment will then mostly accommodate itself to the new order – an easier trick, for most partisans, than flipping to the other side of the aisle entirely.

In other democracies with different political systems, different sorts of fractionalization develop. In many continental democracies, where proportional representation encourages the formation of multiple parties, the political realm has fractured into an unmanageable tangle of parties, which makes the formation of stable governments difficult (and threatens, in some cases, to bring dangerously radical governments into the governing coalition). Separatist and nationalist parties are on the rise across Europe. Le Pen's National Front is ascendant under the leadership of his daughter, Marine Le Pen. Radical parties in Hungary and Poland are pushing for significant political change: to undermine existing democratic institutions and to edge away from the EU.

Even in Westminster systems, such as the original in Britain, which are meant to encourage two-party elections, polarization of interests has given way to partisan splits. As of 2016, the normal Labour–Tory divide (which is challenged in places by the Liberal Democrats) is joined by a separatist Scottish National Party, whose stated aim is to remove Scotland from the United Kingdom, and also by the UK Independence Party, which would like to remove Britain from the European Union (an ambition which looks closer to realization than ever after a referendum held in June 2016 delivered a vote in favour of leaving the EU).

These developments can be seen as part of a second phase of ideological reshuffling (the first being the ideological awakening of groups within existing parties, leading to polarization). The third will be the competition of these new parties (and newly ideologically radical parties, in America) in the electoral arena, leading, in some cases, to fundamental changes in the political stance of major democracies. In the fourth phase, the governments generated by that competition will interact with each other, and with international institutions like the EU, in unpredictable ways. The eurozone, for example, has so far proven remarkably durable in the face of economic catastrophe. On the other hand, it has spent a quarter of its short life in crisis, and only one of the currency union's members needs to elect a government determined to exit to seriously, and perhaps fatally, undermine the entire project.

A COMPETITION OF VISIONS

Radical factions and parties battling for supremacy each have at their heart a particular conception of the 'good life'. Writers and thinkers, like me, try to imagine these post-work utopias, in which, for example, sensibly structured social safety nets could free people of the constraints of the typical job. These people could then offer their services by the hour or the job on new-fangled market-making apps, among other things, or they could even abandon labour markets altogether, as new forms of social institution encouraged them to volunteer their time to the community or otherwise engage in pro-social behaviour – while also living alongside people from vastly different backgrounds and perhaps nationalities, if some of us get our way.

But that is possibly not what the typical rich-world citizen would consider the 'good life', however much we might want that to be the case. We should instead anticipate that voters in many countries, rich and poor alike, will want something more predictable than life governed by supply-and-demand matching apps; more structured than life on the perpetual dole; more comfortable and familiar than life surrounded by people who do things in different ways, speak different languages, and worship different deities.

Indeed, in thinking about conceptions of the 'good life', it is worth considering the life that the working rich have made for themselves. Most live in just a few cities, in nice neighbourhoods surrounded by others very much like themselves: well-off, professionally ambitious, with an interest in metropolitan amenities, and sharing a similar set of values concerning the importance of work, friends and family. They labour long hours, but at jobs that are generally challenging and satisfying, in which their contributions to and stakes in the business are often very clear. Social ties within the group are surprisingly strong: they network with each other, join neighbourhood and community groups, and become involved in civic politics.

Theirs is a life of comfort, but also of purpose and of community.

Most people would, of course, value more leisure time. Certainly, those now working at very unpleasant jobs simply to make ends meet would be glad to be free of the need to work in such fashion. In any future in which technology frees workers of the need to spend most of their daytime hours on the job, many people will opt for much more down time, often spent in rather aimless fashion. (Survey data suggest that, over the last decade, people saddled with extra free time thanks to weak job markets spent much of it sleeping and watching television.)[6]

Yet people of all backgrounds also seem to value narratives of personal ambition and responsibility. People wish to have control over their economic lives and to be seen as contributing both to society and to the well-being of their families. People desire *agency*. They do not wish to be forced into unpleasant work by the need to feed their families, but neither do they want to be written off as unnecessary – or assigned meaningless work as the price of a generous welfare cheque.

It isn't clear that the digital economy can provide the working conditions needed to extend the possibility of bourgeois comfort and status to a broader class of people. That will not stop them desiring it.

The conflict between what people want and what economic and political systems are able to provide will play out in the political arena. Political battles will increasingly feature narratives about how to restore us all to a world in which people work at purposeful jobs for good pay. Those narratives will be thick with bogeymen: the malevolent forces denying voters access to that 'good life'. Conniving foreign governments, job-stealing immigrants, greedy bankers and incompetent politicians all star in such roles.

Demagoguery can be a compelling political force. But to win over the median voter, politicians will probably need to offer a plausible explanation for what has happened to the 'good life' and what steps can reasonably be taken to restore it. Reformers can compete in this arena. There will be room for leaders willing to say that the 'good life' of misty memory cannot be brought back; who promise instead to push forward modest, incrementalist policies, such as those favoured in the era of moderation.

The difficulty the reformers will face is that the global economy will tend to punish such effort. Labour abundance and structural demand weakness are not the sorts of things national politicians working in isolation can fix. They can ameliorate the worst effects, of course, but that will leave voters disappointed, which is what they have been for most of the last few decades.

Moderate reformers will find themselves losing ground to politicians keen to unpick elements of the era of moderation, from the move towards freer trade and capital flows to the elimination of labour-market protections. Politicians will promise to make markets create good jobs: by mandating higher minimum wages, supporting occupational certification and other job protections, and pushing firms to regularize work in sharing-economy sectors – by requiring payment of benefits and the guarantee of a certain number of regular hours, for instance.

The global economy probably won't reward *these* efforts either. But they benefit from securing the support of portions of the electorate who receive protections from such measures – whose slice of the pie is cut a bit larger. In a world in which the coalitions of interests that supported globalization are breaking down, the politics of protection could prove newly durable.

Political tastes don't translate smoothly into policy action. They are expressed in political battles. But the outcomes of those battles depend on power. Labour bargaining power within many economies is at a 100-year nadir. Today's labour victories, when they occur, tend to come from straightforward issues for which it is easy to muster broad, passionate electoral support: policies such as a rise in the minimum wage or a reduction in immigration. The more complex negotiations that occurred a generation or two ago, when labour had a seat at the political table, tend not to occur any longer.

That could change. Drivers for car-sharing firms, such as Uber and Lyft, are battling to unionize. Unionization could eventually come to other sectors of the economy in which large pools of on-demand

labour sell their time through market-making apps as well. Unionization would yield uncertain direct benefits to workers within these
firms, though. Short-run concessions wrung from ownership might
simply accelerate the pace of automation: troublesome labour tends to
encourage the deployment of robots, whether the setting is a factory
in Shenzhen or a car on California streets.

If unions persist long enough and appear often enough, however,
they could begin to cooperate with each other, to strike deals to support
each other's political priorities. In labour organizers' dreams, that
cooperation becomes a class-consciousness – a solidarity – which
would provide a coherence to a labour political agenda. In a world of
diffuse, unorganized labour, policies which boost the fortunes of one
small group of workers (occupational protections for hairdressers,
for example) at the expense of other workers (who then pay more for
their haircuts) can be vulnerable to appeals to a self-interested majority, that such policies are inefficient and bad for most workers. In a
political world of solidarity, such appeals would be less effective;
groups of workers would instead support each other's efforts to capture more income in turn.

While unions are on the march in places, the future for organized
labour does not strike me as especially bright. In the nineteenth
century, firms massed labour together in large factories and cities,
encouraging coordination and strengthening the ability of the workers
to extract concessions from ownership by acting en masse. Occupational workforces today are far more diffuse, and there are more
technological tools available to firms to undermine labour power.

From a narrow economist's point of view, the poor outlook for
organized labour is a good thing on the whole: markets work better
when there is free entry. In practice, the efficiency gains achieved by
fluid labour markets have not been redistributed to the workers
whose bargaining power was sacrificed to achieve that efficiency. Just
as importantly, the absence of a coherent labour political bloc means
that voters motivated by economic and cultural angst will be more
susceptible to demagoguery with mass appeal. Free-floating anger, or
even free-floating dissatisfaction, is not a pleasant thing to have washing around a population.

AGEING WON'T SAVE US

It is worth briefly mentioning the issue of population ageing. A country in which a large share of the population is beyond the typical retirement age could easily have very different political priorities than one in which the bulk of the population is in its prime working years. And indeed, there is some evidence to suggest that older countries are more supportive of immigration than are younger ones. Not only are retirees no longer in competition with immigrant labour, cheap labour also reduces the cost of the medical and care services they disproportionately use. One could just about imagine a future in which an ageing rich world welcomes in lots of young workers from poor countries – to work as help in the home, as physical therapists, as nurses, to help pay the pension bill of the older generation – then bequeaths to those workers and their children the strong rich-world economic and political institutions, to enjoy and make their own.

That would not be an unattractive model, but how realistic is it? Certainly, in practice, ageing countries in Asia and eastern Europe have not been keen to welcome in a flood of new, foreign labour. To some extent, the economic openness of older generations to immigration is dampened by a tendency to cultural conservatism in older countries. What's more, ageing countries are not uniformly old; even in places with highly top-heavy population pyramids, a large share of the population is still of working age and is bound to be resentful of large numbers of people brought in *expressly* to fill jobs in one of the few sectors that reliably creates new employment.

An openness and cosmopolitanism inspired by demographic change would be an encouraging political development. It might one day materialize. It hasn't yet.

THE SHARING ECONOMIES

Could there be a constituency for a more benign set of policy innovations: for generous basic incomes tied to sensible work requirements, designed to encourage public-spirited labour contributions but leaving room for the individual's freedom to live the way he or she wants to live? That might be a lot to ask. But we might expect generous welfare policy to emerge in places where the solidarity that appear is community-based, rather than class-based. Unfortunately, that will tend to occur within ethnically or nationally coherent political units. It is no wonder that experimental, generous welfare policy has tended to emerge in Nordic countries, where ethnic and communal ties are strong (but where openness to immigration has begun to tear at the social consensus).

Indeed, as the politics of the digital era evolve, two geopolitical forces tend to push against each other by turns. The first, which is especially apparent now, is a tendency towards fractionalization of existing states into smaller chunks. Hyperglobalization means that even very small economies can enjoy access to global markets, which reduces the advantage of being part of a much larger state. Within the superstate of the European Union, separatism is especially attractive: provided an enclave can maintain access to the EU market, separation enables greater local autonomy and greater within-group redistribution. Scots can share their riches with Scots*and Catalans with Catalans, without the interference of or without needing to extend redistribution to other, out-groups.

What separatist quasi-nations seem to want is a world in which they enjoy the economic benefits of global integration, but in which critical political and economic decisions are made by units with a high degree of national or ethnic coherence: a future of Irelands and Estonias rather than of Britains and Spains: larger states with more diverse populations.[7]

It isn't clear that new examples of this model can appear without doing irreparable damage to the broader, integrated market. The institutions of the EU are not built to handle waves of fracturing nations. Italian, Belgian and even German leaders are understandably reluctant to sign off on Catalan independence, given the damage regional separatism could do to their own states. Rich-world

---

* Scottish wealth for Scots looks slightly less attractive with oil at $30 per barrel than it did when oil was at $100 per barrel.

ethno-nationalism could destroy the economic integration on which its prosperity depends.

If it doesn't fail before it begins – if separatism does achieve some successes – the model of highly redistributive, ethno-nationalist mini-states participating in an open global economy might nonetheless prove unsustainable. Rich places with generous welfare states are desirable countries to live in: people will seek to migrate to such places. Those places can then either allow migrants in, undermining the ethno-nationalist coherence that enables redistribution, or shut them out, undermining the economic integration that enables prosperity. Looking at the political evolution of the European project, one might suppose that the increased ethno-nationalist consciousness nurtured by economic integration is inconsistent with the sustaining and deepening of that integration. The former seems to contain the seeds of destruction of the latter.

And as that progression plays out, a second geopolitical force might then assert itself: the safety of bigness. Very large states are attractive when international economic integration breaks down and when national security seems to be threatened. Big countries have big internal markets and are capable of supporting big, powerful militaries. They thrive when international relations and economic integration break down.

The American experience, however, suggests that big markets with heterogeneous populations struggle to support high levels of internal redistribution. America is a big, successful melting pot. The ethno-nationalist diversity of the American population, however, has long been an obstacle to the construction of an exceptionally generous welfare state. White voters in the South are sceptical of a welfare state that promises to deliver generous support to black Americans in northern cities, or to Latin Americans in California.

Big, diverse nations contain lots of communities of affinity: groups that feel more like themselves than like others. Communities of affinity are natural locuses for redistribution, but redistribution within those communities can only occur when such communities line up, more or less, with the apparatus of the state. Communities of affinity will therefore try to shrink the boundaries of the state down to fit them. Sometimes, on the other hand, external forces – such as war or economic crisis – will conspire to broaden communities of affinity: to temporarily reduce the salience of one's ethnicity relative to one's nationality.

Large, ethnically heterogeneous states, such as America and the bigger European countries, were able to build inclusive economies with healthy levels of redistribution in the immediate post-war era. Yet that example is hardly encouraging. Then, the most salient community of affinity was the state, perhaps even the West as a whole, which was pitted in a struggle for survival against Communism.

Some time in the future, a wonderful new politics might well emerge which provides a robust minimum standard of living to all regardless of race or nationality, which supports a multitude of different conceptions of the 'good life', and which does not rely on some underlying fear of some outside other to maintain its popularity. We are not yet able to conceive of such a system, or to understand what balance of political forces needs to emerge to bring it into existence and sustain it. And so, for the time being, we are stuck in a world of nasty political trade-offs. States will attempt to shrink themselves to a level of homogeneity conducive to redistribution; or they will stay large and non-redistributive and unequal and vulnerable to the passions of demagogues; or they will stay large and become communitarian and redistributive thanks to the strain of outside geopolitical pressures.

We can but hope this era will prove a fleeting one. History suggests it will not be. But perhaps we will get lucky.

PHANTOM INSTITUTIONS

Perhaps we will get lucky. Perhaps we will not. Times of change, like the present, are dangerous. There is no sense in regretting the danger or trying to wish it away; the arrow of history cannot be made to point in the other direction. But it is worth pointing out the danger in order to encourage those with a mind to do so to work hard to create the best possible future.

On 13 February 2016, Antonin Scalia, justice of the Supreme

Court of the United States, died at the age of seventy-nine. Immediately, leaders of America's two political parties swung into action, planning and strategizing over the battle to confirm his replacement. It quickly became clear that Republican leaders were interested in using the legal, procedural tools available to them to block any nominee to the court put forward by Democratic president Barack Obama – and would step outside the bounds of Congressional norms if necessary, even if such action precipitated something like a constitutional crisis.

Jonathan Chait, a left-leaning writer at *New York Magazine*, observed the spectacle and wrote, 'It turns out that what has held together American government is less the elaborate rules hammered out by the guys in the wigs in 1789 than a series of social norms that have begun to disintegrate.'[8]

In fact, the rules hammered out by the guys in wigs never held together the American government. Those rules have no agency, no ability to express a view, and no army to command. There is no action that one can take that will move the American Constitution to animate itself and discharge retribution.

The Constitution has only ever had power because people behaved as though it did. The Constitution merely expresses some of the norms by which participants in American government behave: it is the embodiment of some of the most important elements of American social capital.

The Founders recognized that the social capital of the American republic could only be powerful, and could only be a means through which to realize their dream of a new kind of state, if respect for and adherence to the norms in the Constitution became a key part of the cognitive hardware of America's governing elite. And early in the republic's existence its leaders checked their own behaviour when they saw it as being in conflict with the norms they had set out in the Constitution. President James Monroe shocked his colleagues and advisers when he vetoed a bill providing for public infrastructure investment – a policy he supported – because he determined that it was unconstitutional. It was actions such as these which built up the Constitution into something more than ink on parchment: an entity of its own, in the eyes of men and women, wielding its own power.

But when an institution achieves that sort of status, people begin to forget that it is only the determined action of individuals behaving according to the social capital that lives in their heads that holds society together. Like a deity or a parent, the rules of the institution become things to be subverted when the opportunity presents itself. When people see themselves as independent individuals living under the authority of an external code, rather than as participants in a social consensus, they see little reason not to test the boundaries of the imagined external authority. But because they are, in fact, participants in a social consensus, those actions chip away the power of the social capital, undermining it bit by bit, until the value of adherence to the old social norms disappears and takes with it the once-venerated institution.

The American Constitution, we can hope and I certainly expect, is not about to suffer such a fate. But the modern world is built of many overlapping institutions, which represent many different forms of social capital. When they erode, they can be difficult to repair, and the failure of some critical institutions can trigger the collapse of others that depend upon them. Democracy, tolerance, liberalism, respect for individual autonomy: these are all norms that residents of rich countries have so deeply internalized that they often fail to realize when such institutions are in need of serious defence.

'I like to pay taxes,' Oliver Wendell Holmes, another Supreme Court justice, is thought to have said. 'With them I buy civilization.'[9] The global market economy is a force for the creation of mass wealth. Living within its size and power and complexity, it is easy to forget that there is no independent entity called the global economy. There are only people, operating according to the social capital they carry in their heads. The global economy, just like any other institution built of nothing more than social consensus, can be weakened by those who seek to subvert its norms. It can be hobbled or destroyed. People who behave as though the market economy is an immutable thing, and who take as much as they can, believing that the system can and will thrive without cooperation to keep the social consensus in favour of it together? Well, they take rich societies closer to a world in which everyone is made worse off.

The only way forwards is through broad social agreement that what we are doing is better than alternative paths. If we don't all work hard to build an agreement as encompassing and as broadly enriching as possible, then the avenues for social agreement grow narrower, and the world becomes a more fractured, a more impoverished, and a more unhappy place.

12

Human Wealth

I am a baseball fan. In October of 2012, the Washington Nationals, my team, made the playoffs for the first time in the team's history. The first round was a five-game series against the Saint Louis Cardinals. The Nats fell behind in the series, two games to one, then won the fourth game – a long, tense pitcher's duel – in epic fashion. In the bottom of the ninth inning, Jayson Werth, one of the team's stars, drove a scorching line drive into the stands for a game-winning home run. Nationals Park erupted in celebration. My two-year old daughter joined me in dancing and cheering in front of our television at home. The series was tied at two games to two. I had tickets to the decisive game five.

Walking into the park that chilly evening, there was a feeling of giddiness that rippled across the crowds streaming through the gates. The fans squeezed into the stadium were exuberant as the Nats built a three-run lead in the first inning, then shouted and sang as a pair of home runs in the third inning increased the lead to six runs. Then, in excruciating fashion, the Cardinals clawed their way back into the game, scoring a run here, a run there, until, by the top of the ninth inning, they trailed by just two runs, seven to five. The stomachs of the assembled masses knotted in unison. Twice Cardinals batters were just one strike away from defeat. But their hitters came through: the Cards scored four runs in the ninth and won the game nine to seven. It was heartbreaking. Nationals fans trudged mournfully out of the park. But, in that weird sports way, the collective nature of the sadness was oddly thrilling. Our collective disappointment would become part of the team's collective memory, the sense of shared narrative that helps fans explain why they feel so strongly about one particular team, and, indeed, a part of the community that makes the team what it is and helps it to succeed.

The value generated by Major League Baseball is collective in nature. The game would not be the game without the players, who dedicate themselves to their craft and whose tireless efforts create the spectacle that is the game of professional baseball. But there would be no reason for players to devote themselves so single-mindedly to the game if there were not the possibility of fame and financial reward at the end of the process, and the reward at the end of the process would not be there were there not millions of fans willing to fill the stands, and watch the games on television, and buy the merchandise. The billion-dollar team valuations enjoyed by club owners would not be possible without the efforts of the players and the passion of the fans.

Baseball is a good metaphor for most things in life, and the economy is no exception. As in baseball, value is fundamentally social in nature: it is the collective passion and interest that makes the sport such a valuable institution. As in baseball, things such as productivity and scarcity shape the distribution of rewards in society. As in baseball, bargaining power is of critical importance in determining the distribution of rewards. As in baseball, abuse of bargaining power can reasonably be called unfair, and enough abuse can precipitate a social reaction that threatens the fundamental value of the enterprise as a whole – such as when the strike of 1994 gutted fan interest and cost baseball its position as America's national pastime.

As in baseball, it is easy for all the participants in the economy to convince themselves that their participation is what matters, that they are the authentic creators of value, that their effort is what ought to be rewarded most handsomely. And everyone has a point. But while we can rely on economics to do some of the work of sorting out who deserves what, we are kidding ourselves if we think the invisible hand can be entrusted to handle the whole job. Left alone, the invisible hand is simply the thudding fist of the powerful. It would be wonderful if things were otherwise, but they aren't.

Like most people, I often wonder if I am paid fairly. I like to think that I am very good at my job. But I am keenly aware of my bargaining power. It amounts to this: I can threaten to go and maybe *The Economist* won't want me to. And yet there are large numbers of people out there who could do my job well. My training and fluency in *Economist* culture are important to my professional success, and yet there are vast numbers of people out there with similar skills and experience, who could learn the culture if given the opportunity.

Of course, I work very hard. I put in long hours. To some small degree, I work long hours because that extra effort and contribution boosts the firm's bottom line, and a bigger bottom line means a little more money, thanks to *The Economist*'s profit-sharing programme. But the link there is too small to have anything but the tiniest effect on my compensation.

The main reason I work hard is because the value within *The Economist* is social, and that social value is distributed over a limited set of economics writer positions, and I mean to cement myself in one of those positions. And the way to accomplish that is to distinguish myself and to create the general impression of indispensability. The hours I invest are a critical part of the case I make to my employers to put me in my job in the first place. And it is the job that is the prize, because the value generated by the firm is so overwhelmingly social in nature: our culture and collective knowledge are our competitive edge; the whole is so much greater than the sum of the parts.

The wealth of humanity is limited by our ability to produce goods and services of value. The production of goods and services of value increasingly rests on the collection, processing and management of information. There is no value without the knowledge of what can be produced, what ought to be produced, and how it can be produced most effectively. It is the information-processing structures of firms, cities, nations, and other institutions of human society that gather that information, and sort it, and turn it into the production that enriches people around the world. The wealth of humans is societal. But the distribution of that wealth doesn't rest on markets or on social perceptions of who deserves what but on the ability of the powerful to use their power to retain whatever of the value society generates that they can.

That is not a radical statement. People take what they can take, and it is only the interplay of countervailing forces and the tolerance of the masses that limits that impulse – that works to create institutions that limit that impulse.

It is impossible to imagine Bill Gates's wealth without Bill Gates's ingenuity and effort. But it is far easier to imagine Bill Gates's wealth being produced by someone other than Bill Gates within the institutions of modern American economic society than it is to imagine Bill Gates generating Bill Gates's wealth in a different time and place – in France in the 1700s, or in the Central African Republic today – in which society was or is less tolerant of entrepreneurial capitalism and the accumulation of personal billions, and where the community of engineers that gave rise to and became America's tech sector is absent. Indeed, at some point in Microsoft's history it was Microsoft the information-processing organism that was more critical to Bill Gates's wealth accumulation than Bill Gates himself. People, essentially, do not create their own fortunes. They inherit them, come to them through the occupation of some state-protected niche, or, if they are very brilliant and very lucky, through infusing a particular group of men and women with the germ of an idea, which, in time and with just the right environment, allows that group to evolve into an organism suited to the creation of economic value, a very large chunk of which the founder can then capture for himself.

WE CAN DO BETTER

In a way, it would be much easier if the robots were simply taking all the jobs. Solutions might not be any more straightforward to come by, but the sight of millions of robot dog-walkers and sanitation workers strutting through crowds of unemployed humans would at least be clarifying.

Instead, the remarkable technological progress of the digital age is refracted through industrial institutions in ways that obscure what is causing what. New technologies *do* contain the potential to revolutionize society and the economy. New firms *are* appearing which promise to move society along this revolutionary path. And collateral damage, in the form of collapsing firms and sacked workers, is accumulating.

But the institutions we have available, and which have served us well these last two centuries, are working to take the capital and labour that has been made redundant and reuse it elsewhere. Workers, needing money to live, seek work, and accept pay cuts when they absolutely must. Lower wages make it attractive for firms to use workers at less productive tasks. The flow of people into low-productivity work has had the effect of making society look poorer than it is. And low wages have also made society poorer than it ought to be: by making it more difficult for governments to manage the economy, and by reducing the incentive to invest in labour-saving technology.

This process will not end without a dramatic and unexpected shift in the nature of technology, or in the nature of economic institutions. Changes in technology are hard to predict, but as technological capabilities improve, the set of tasks at which humans retain an advantage shrinks. Changes in economic institutions are a little easier to reason through. Because productive societies, and especially nation-states, are the locus of redistribution, both the productive and unproductive workers within them have an interest in drawing the border of the group as tightly as possible. It is the fact of redistribution that leads society to prioritize the effect of *scarcity* on the sharing of gains within society over the effect of the society itself on the *productivity* and *welfare* of those who are allowed to enter.

That is no argument for abandonment of redistribution: unless technological change dramatically alters the demand for human labour in a way that seems both unlikely and which has occurred very rarely in industrial history, redistribution of one sort or another *is the way* that the incomes of less-productive workers are made to keep up with growth in average output per person. What's more, less-productive workers have a right to redistribution, both because an excessive imbalance of incomes is, or ought to be, an affront to our sense of economic justice – hard work and ingenuity should be rewarded handsomely; the blind luck of being born talented in a productive, market-oriented country should not – and because all members of a society contribute, in ways we can't always perceive, to its sustainability.

It is instead a call to recognize that this current state of the world is an absurdity. The point of technological progress, if there possibly is one, is to improve human lives: to make as many people as possible as well off as possible. Is there any reasonable story available which explains how it is that poverty in developing countries, or in the ghettos of disadvantage in rich countries, is a necessary part of the system that provides us with smartphones and luxury cars and enriches a relative handful of executives and financiers? Is it really the case that the one can't be got rid of without threatening the system that provides for us the other?

Of course not. The worst inequities of industrial history were never a necessary accompaniment to the march towards greater prosperity. Troublingly, impressive recent advances in technology do not seem to be bringing us any closer to grappling with this absurdity. The better we get at making things, the more bizarre the distribution of income looks. Technological change *has* enabled growth in living standards in the emerging world – thanks to its extension of the market system, rather than any burst of humanitarian empathy. But that growth remains mostly incomplete, and it has come alongside stagnation in conditions for many rich-world households. Indeed, it is fashionable for haves to muse that we ought not to worry too much about the struggles of the rich world's have-nots, given income growth among the emerging world's most fortunate.

That hardly seems like an effective argument to build long-term support for the status quo. And the status quo, when it changes, will be pushed in the direction of increased social distance: the use of law and custom to try to push open gaps between societies where technology is closing them, sought because existing social structures are failing to transform new economic possibilities into broad-based income growth.

THE WEALTH OF HUMANS

In his *Inquiry into the Nature and Causes of the Wealth of Nations*, Adam Smith mused on the way in which market economies translate human impulses into social wealth:

> [M]an has almost constant occasion for the help of his brethren, and
> it is in vain for him to expect it from their benevolence only. He will

be more likely to prevail if he can interest their self-love in his favour, and show them that it is for their own advantage to do for him what he requires of them. Whoever offers to another a bargain of any kind, proposes to do this. Give me that which I want, and you shall have this which you want, is the meaning of every such offer . . . It is not from the benevolence of the butcher, the brewer, or the baker, that we expect our dinner, but from their regard to their own interest. We address ourselves, not to their humanity but to their self-love, and never talk to them of our own necessities but of their advantages.[1]

Smith was seeking to replace one view of the way the wealth of the world is generated with another. The prevailing view at the time, in the latter half of the eighteenth century, was that large trade surpluses were the route to riches – the larger a country's surplus, the greater the inflows of gold and silver – which implied that riches were zero-sum in nature. A larger surplus for one country necessarily required a smaller surplus for another. This antagonistic, 'mercantilist' world encouraged a worldview sympathetic to imperialism and war.

Smith saw things differently. Trade is not zero-sum, he wrote. Rather, trade increases the size of the market, which allows for greater labour specialization. Specialized labour is more productive than non-specialized labour, so that a world of trade and specialization, in which many people focus on one task and exchange their produce with others in mutually beneficial trades, is one in which everyone is much better off than a world in which individual countries seek to buy as little as possible from competitors. The 'common wealth' is maximized when people are left free to follow their self-interest and exchange whenever and with whomever they choose.

It is a beautiful and important intellectual model of the world. But it is incomplete. Self-interest governs more than our behaviour in labour and product markets. It also governs our attitudes and behaviours towards the societies in which we belong. Societal openness generates broad benefits but localized costs. And so people rationally seek to limit societal openness, out of self-interest.

But if the locus of redistribution could be changed, then the zero-sum aspect of societal openness could be defused. Secure in the knowledge that societal growth would not reduce redistribution (and could indeed increase the value available for redistribution by increasing global output) the incentive to draw the borders of society tightly would be curtailed. The challenge, of course, is to create the broad social interest in an encompassing redistribution. How to do that?

There is the hint of an answer in Smith's other great work, the *Theory of Moral Sentiments*, which opens:

> How selfish soever man may be supposed, there are evidently some principles in his nature, which interest him in the fortune of others, and render their happiness necessary to him, though he derives nothing from it except the pleasure of seeing it. Of this kind is pity or compassion, the emotion which we feel for the misery of others, when we either see it, or are made to conceive it in a very lively manner. That we often derive sorrow from the sorrow of others, is a matter of fact too obvious to require any instances to prove it; for this sentiment, like all the other original passions of human nature, is by no means confined to the virtuous and humane, though they perhaps may feel it with the most exquisite sensibility. The greatest ruffian, the most hardened violator of the laws of society, is not altogether without it.[2]

The force of human empathy can be made to serve either openness or societal mercantilism. The question we ask ourselves, knowingly or not, is: with whom do we want to share society? The easy answer, the habitual answer, the ancient answer is: with those who are *like us*.

But this answer is bound to lead to trouble, because it is arbitrary, and because it is lazy, and because it is imprecise, in ways that invite social division. There is always some trait or characteristic available which can be used to define someone seemingly *like us* as *not like us*.

There is a better answer available: that to be 'like us' is to be *human*. That to be human is to earn the right to share in the wealth generated by the productive social institutions that have evolved and the knowledge that has been generated, to which someone born in a slum in Dhaka is every bit the rightful heir as someone born to great wealth in Palo Alto or Belgravia.

The difficulty we face is *managing the thing*. We must try not to destroy the good institutions we find in front of us, the workings of which we do not entirely understand. In seeking to make the world a better place, we must be cognizant of the fact that this matters, and that we can't reasonably expect even the most empathetic of societies to throw open their borders heedlessly when no other country is doing so, and when the pool of potential migrants dwarfs those living and working within those rich societies.

But we should also realize that those societies do not belong to us. If we are lucky enough to find ourselves within them, we can argue credibly that we are contributing to them and therefore deserve a share of the benefits that flow from them. But the fact that we are lucky enough to be within them and contributing to them does not confer on us the exclusive right to such a position. If anything, it confers on us the responsibility to try to make the society as robust as possible, so that its membership can be extended to as many people as possible. No one deserves to be poor. No one deserves to be arbitrarily rich. Rich societies can find ways to justify their great wealth relative to others: their members can tell themselves stories about the great things they did that others could not have done that made them wealthy beyond imagination. Alternatively, they could recognize the wild contingency of their wealth, cultivate human empathy, and do what they can to extend the wealth of humans to everyone.

It took me a while to realize that not everyone grew up with a great sea of lawn around their childhood home, and that not everyone had the great luxury to grouse about the work their fathers made them do for an hour or two on a Saturday in hopes of teaching them not to take a comfortable life for granted. I did eventually find the will to work long, hard hours, but the fear of discomfort has never been among the more important motivating factors. I have been lucky enough to find myself in a field in which passion, ambition and a sense of healthy professional competition are much more acute motivational sensations.

Even so, it has been hard to take ownership of what good fortune I have enjoyed. I also came to realize that nothing I ever did in my life was as likely to affect my personal material comfort as much as the actions my father took decades ago, when he left his childhood farm to go to college, bidding farewell to rural life for a career as a professional in a growing metropolis. But then nor could he take credit for being born white, male and American.

In another age, more of those who grew up around him might have found their way to better jobs and lives, working, perhaps, in the local textile mill that once employed my grandmother – if they could bear the heat and ear-splitting noise, and save a bit of the meagre hourly wage. But the mill is gone, a victim of trade and technology and time. There is much to be said for climbing economic ladders, but it is impossible to climb a ladder that isn't there.

When I return to my childhood home now, I am occasionally there to see how the lawn is taken care of. Once a week a landscaping service team swings by. Two men hop out of a truck. One mounts a massive ride-on lawnmower that races around the property like a go-cart; the other runs a 'weedeater' on a wheel up and down the driveway, then dons an industrial powered leafblower backpack and sends whatever yard waste happens to be lying around flying off in a hurry. They are done in ten minutes. Teams of men just like those who work my parents' lawn operate all over the city. The men, many of them recent immigrants from Latin America, don't earn very much, but most are no doubt grateful for the work. The firm that employs them is a client of my father's accounting firm. It is a successful enterprise.

While I was writing this book, iRobot, the maker of the adorable autonomous vacuuming robot called the Roomba, received regulatory approval for a lawn-mowing version of the tiny hoovering droid. If the mower bot is very successful, it will put many lawn-care crews out of business. If it is only about as successful as the Roomba, it will save some people some time mowing their own lawns while many more will continue to employ people to do the work, just as many households continue to hire cleaning crews to vacuum their floors.

If the bot is a hit, producers of autonomous mowers will make a lot of money, firms that manage lawn-care crews will struggle, and both workers who rely on jobs cutting grass for income and parents who rely on mowing as a source of chores for children will face a dilemma. I might buy one, if I ever have much of a lawn worth fussing about, for the fun of having it, or I might just encourage my kids to help out in the garden for an hour or two each week. I won't ask them to underbid the robot; I'll do my best to keep them as comfortable and happy as I can. Whether within a family or the world at large, it is fair that society should ask for a contribution from its members. If we are clever enough to think up grass-mowing machines, we should also be clever enough and moral enough to maintain social order without threatening members with impoverishment.

Epilogue

This book no doubt comes across as rather gloomy in parts, but it is decidedly optimistic in one sense, which is the belief in the capacity of humanity to develop new and important technologies, and to find ways to use them to improve lives. The digital revolution will prove to be as powerful and transformative as the most fundamental innovations of the industrial age. And that power is potential: the potential to create a mass prosperity of an unprecedented nature.

In assessing how optimistic or pessimistic one ought to be about this possibility, it is worth imagining a person, chosen at random from among those alive in 1850, and describing to them how world events would unfold over the next 150 years. Should that randomly chosen person have been optimistic about the technological and economic changes underway?

Sadly, the answer is ambiguous. Among those alive in the decades after 1850, some individuals enjoyed historically unprecedented increases in economic opportunity. Most others did not. Of those alive in 1850, some produced distant descendants who, more than a century later, enjoyed incomes and life expectancies and experiences beyond the imagination of the greatest science fiction writers of the era. Life was better, immeasurably in some cases, in almost every way. Those descendants, though, could reflect on what a close scrape the journey had been, with great wars and depressions in the intervening period, culminating in a nuclear-tipped stand-off between economic ideologies that nearly destroyed all of humanity.

And of those alive in 1850, whose descendants survived through to the late twentieth century, most parented generations of people whose lives improved very slowly, very unreliably, very incrementally, right through the end of the twentieth century – when the average real income in sub-Saharan Africa was roughly that enjoyed by Britain in 1800.

Average incomes did improve though, and they might have improved more, given more sensible policies from those who enjoyed the best that technology and that social capital had to offer. The best reason for optimism now is that humanity has the experience of the industrial revolution under its belt. It has been through that wrenching transformation, seen its dangers, and understood the ways in which it was eventually made to improve lives on a broad scale.

The reason to be pessimistic is that now, as in the industrial era, there is no one in control. There is no navigator with a map of the past in hand who can judiciously pilot modern society towards a world in which technology is empowered to generate the greatest good for the greatest number of people. The reason to be pessimistic – or, more appropriately, the reason to be both realistic and actively idealistic – is that the only way society advances is through the chaotic, haphazard and wild interaction of social forces of all sorts. And there is no way to be sure it will conclude as propitiously this time as it did the last.

We are entering into a great historical unknown. In all probability, humanity will emerge on the other side, some decades hence, in a world in which people are vastly richer and happier than they are now. With some probability, small but positive, we will not make it at all, or we will arrive on the other side poorer and more miserable. That assessment is not optimism or pessimism. It is just the way things are.

Face to face with the unknown, it is hard to know what to feel or what to do. It is tempting to be afraid. But, faced with this great, powerful, transformative force, we shouldn't be frightened. We should be generous. We should be as generous as we can be.

Notes

Abbreviations

BEA	*US Bureau of Economic Analysis*
BLS	*US Bureau of Labor Statistics*
EIA	*US Energy Information Administration*
IMF	*International Monetary Fund*
NBER	*National Bureau of Economic Research*
OECD	*Organization for Economic Co-operation and Development*

EPIGRAPH

1 Smith, Adam, *An Inquiry into the Nature and Causes of the Wealth of Nations* (London: W. Strahan and T. Cadell, 1776).
2 Keynes, John Maynard, 'Economic Possibilities for our Grandchildren', *Essays in Persuasion* (London: Macmillan, 1931).

INTRODUCTION

1 'The Onrushing Wave', *The Economist*, 18 January 2014.
2 'Is 4.4 jolt an end to Los Angeles' "earthquake drought"?', *Los Angeles Times*, 17 March 2014.
3 http://www.statista.com/chart/612/newspaper-advertising-revenue-from-1950-to-2012/
4 OECD, Average Annual Wages, in national currency units, constant (https://stats.oecd.org/Index.aspx? DataSetCode=AV_AN_WAGE).
5 Karabarbounis, Loukas, and Neiman, Brent, 'The Global Decline of the Labor Share', *Quarterly Journal of Economics*, June 2013.
6 World Top Incomes Database (http://www.wid.world/#Database).
7 BLS, Current Population Survey.

8   Case, Anne, and Deaton, Angus, 'Rising Morbidity and Mortality in Midlife Among White Non-Hispanic Americans in the 21st Century', Woodrow Wilson School of Public and International Affairs and Department of Economics, Princeton University, Princeton, NJ, 08544, 17 September 2015.

9   Eurostat, Labour Force Survey.

10 Piketty, Thomas, *Capital in the Twenty-First Century* (Cambridge, MA: Harvard University Press, 2014), first published in French as *Le Capital au XXIe siècle* (Paris: Editions du Seuil, 2013).

11 Dobbs, Richard, Madgavkar, Anu, Barton, Dominic, Labaye, Eric, Manyika, James, Roxburgh, Charles, Lund, Susan, and Madhav, Siddarth, 'The World at Work: Jobs, Pay and Skills for 3.5 Billion People', McKinsey Global Institute, June 2012.

12 Jean-Baptiste Say (1767–1832), French economist and businessman, best known for Say's Law.

13 Mason, Paul, *Postcapitalism: A Guide to Our Future* (London: Allen Lane, 2015).

14 Keynes, 'Economic Possibilities'.

15 Ibid.

16 Maddison Project Database.

17 Milton Friedman (1912–2006) was an economist – arguably the second-most consequential economist of the twentieth century after Keynes. His most important work, with Anna Schwartz (1915–2012), *A Monetary History of the United States, 1867–1960* (Princeton, NJ: Princeton University Press, 1963), explained the critical importance of monetary policy to economic stability (and blamed central banks for allowing the Depression to become so bad). Yet he was ultimately best known in the popular imagination as the leader of the small-government vanguard of economists and policy wonks which helped bring about the Reaganite and Thatcherite reforms of the 1970s and 80s.

18 Maddison Project Database (http://www.ggdc.net/maddison-project/data.htm).

19 Brynjolfsson, Erik, and McAfee, Andrew, *Race Against the Machine: How the Digital Revolution is Accelerating Innovation, Driving Productivity, and Irreversibly Transforming Employment and the Economy* (Digital Frontier Press, 2011).

20 Ford, Martin, *Rise of the Robots: Technology and the Threat of a Jobless Future* (London: Oneworld Publications, 2015).

21 Hayes, Christopher, *Twilight of the Elites: America After Meritocracy* (New York, NY: Crown Publishing Group, 2012).

22 Marc Andreessen (1971–) helped write the code for Mosaic, an early and important web browser, and co-founded Netscape. He later co-founded Andreessen Horowitz, an important Silicon Valley venture capital firm. Many know him best, however, as one of Twitter's top digital economy gurus.

23 Cowen, Tyler, *Average is Over: Powering America Beyond the Age of the Great Stagnation* (New York, NY: E. P. Dutton & Co Inc., 2013).

24 Mankiw, Gregory, 'Yes, the Wealthy Can Be Deserving', *The New York Times*, 6 February 2014.

25 Corn, David, 'Romney Tells Millionaire Donors What He Really Thinks of Obama Voters', www.motherjones.com, 17 September 2012.

26 On a PPP-adjusted, per capita basis; data from the IMF.

27 Ocean Tomo, 'Annual Study of Intangible Asset Market Value', LLC, 2015.

28 Weil, David, *The Fissured Workplace: Why Work Became So Bad and What Can be Done to Improve It* (Cambridge, MA: Harvard University Press, 2014).

29 US Census Bureau, New Residential Construction.

30 S & P Case-Shiller Home Prices Indexes.

31 'The model minority is losing patience', *The Economist*, 3 October 2015; IMF data.

32 Net migration, United Nations Population Division, World Population Prospects.

1. The General-Purpose Technology

1 Smith, Margaret, ed., *The Letters of Charlotte Brontë: Volume Two, 1848–1851* (Oxford: Clarendon Press, 2000).

2 Larson, Erik, *The Devil in the White City: Murder, Magic and Madness at the Fair that Changed America* (New York, NY: Crown Publishing Group, 2003).

3 US Census Bureau, Population of the 100 Largest Cities and Other Urban Places in the United States: 1790–1990.

4 Gordon, Robert, 'Is US Economic Growth Over? Faltering Innovation Confronts the Six Headwinds', NBER Working Paper 18315, August 2012.

5 Azoulay, Pierre, and Graff-Zivin, Joshua, 'The Production of Scientific Ideas', *NBER Reporter*, 2012.

6 Rosenberg, Nathan, and Trajtenberg, Manuel, 'A General-Purpose Technology at Work: The Corliss Steam Engine in the Late 19th Century US', *Journal of Economic History*, March 2004.

7 Moore, Gordon, 'Cramming More Components Onto Integrated Circuits', *Electronics*, 19 April 1965.

8 Cowen, Tyler, *The Great Stagnation: How America Ate All the Low-Hanging Fruit of Modern History, Got Sick, and Will (Eventually) Feel Better* (New York, NY: E. P. Dutton & Co Inc., 2011).

9 Solow, Robert, 'Manufacturing Matters', *The New York Times*, 12 July 1987.

10 Basu, Susanto, and Fernald, John, 'Information and Communications Technology as a General-Purpose Technology: Evidence from U.S. Industry Data', *German Economic Review*, Vol. 8, Issue 2, 2007.

11 Brynjolfsson and McAfee *Race Against the Machine*.

12 Raymond 'Ray' Kurzweil (1948–) is a serial inventor and entrepreneur whose past innovations include programmes which allow computers to recognize text and convert it to speech. More recently he has become known for his writing on transhumanism, and the prospect that powerful technology will allow humanity to achieve near-immortality.

13 BLS, Current Employment Statistics.

2. Managing the Labour Glut

1 See, for example, the World Top Incomes Database (http://www.wid.world/#Database).

2 'The March of Europe's Little Trumps', *The Economist*, 12 December 2015; 'A Blazing Surprise', *The Economist*, 1 June 2013.

3 Dabla-Norris, Era, Kochhar, Kalpana, Suphaphiphat, Nujin, Ricka, Frantisek, and Tsounta, Evridiki, 'Causes and Consequences of Income Inequality: A Global Perspective', IMF Staff Discussion Note, 1 June 2015.

4 Raff, Daniel, and Summers, Lawrence, 'Did Henry Ford Pay Efficiency Wages?', *Journal of Labor Economics*, Vol. 5, No. 4, Pt 2, October 1987.

5 Meyer, III, Stephen, *The Five Dollar Day: Labor Management and Social Control in the Ford Motor Company, 1908-1921* (New York, NY: SUNY Press, 1981), quoted in ibid.

6 Hall, Jonathan, and Krueger, Alan, 'An Analysis of the Labor Market for Uber's Driver-Partners in the United States', Working Paper, Princeton University, Industrial Relations Section, January 2015.

7 Autor, David, 'The "Task Approach" to Labour Markets: An Over-view', *Journal for Labor Market Research*, January 2013; Frey, Carl Benedikt, and Osborne, Michael, 'The Future of Employment: How Susceptible are Jobs to Computerisation?', 17 September 2013.

8 US Census Bureau, Educational Attainment, CPS Historical Time Series Tables.

9 OECD, Population with Tertiary Education.

10 Abramovitz, Moses, and David, Paul, 'Convergence and Deferred Catch-up: Productivity Leadership and the Waning of American Exceptionalism', from Landau, Ralph, Taylor, Timothy, and Wright, Gavin, eds., *The Mosaic of Economic Growth* (Palo Alto, CA: Stanford University Press, 1995).

11 Katz, Lawrence, and Margo, Robert A., 'Technical Change and the Relative Demand for Skilled Labor: The United States in Historical Perspective', from Boustan, Leah Platt, Frydman, Carola, and Margo, Robert A., eds., *Human Capital in History: The American Record* (Chicago, IL: University of Chicago Press, 2014).

12 OECD, Adult Education Level.

13 James, Jonathan, 'The College Wage Premium', Federal Reserve Bank of Cleveland Economic Commentary, 2012; Autor, David, 'Skills, Education, and the Rise of Earnings Inequality Among the "Other 99 percent"', *Science*, 23 May 2014.

14 Lindley, Joanna, and Machin, Stephen, 'The Rising Postgraduate Wage Premium', *Economica*, March 2016.

15 'The Price of Getting Back to Work', *The Economist*, 1 February 2014.

16 Bosworth, Barry P., 'Sources of Real Wage Stagnation', Brookings Institution, 22 December 2014.

17 US Census Bureau, Income and Poverty.

18 OECD, Average Annual Wages.

19 Bosworth, Barry P., 'Sources of Real Wage Stagnation'.

20 World Top Incomes Database.

21 Nicholas Kaldor (1908–1986) was born Káldor Miklós in Budapest, Hungary, but moved to Germany and then Britain as he studied. In addition to his academic contributions, Kaldor was a long-serving economic adviser to the British government. On 9 July 1974, Kaldor was made a life peer as Baron Kaldor, of Newnham in the city of Cambridge.

22 Karabarbounis, Loukas, and Neiman, Brent, 'The Global Decline of the Labor Share', *Quarterly Journal of Economics*; Elsby, Michael, Hobijn, Bart, and Sahin, Aysegul, 'The Decline of the US Labor Share', Brookings Papers on Economic Activity, Fall 2013.

3. In Search of a Better Sponge

1 EIA (http://www.eia.gov/beta/international).
2 BLS, State and Metro Area Employment, hours and earnings.
3 Logan, Bryan, 'Mercedes-Benz's Self-driving Big-rig Proves that Autonomous Vehicles are Coming Sooner than We Think', *Tech Insider*, 5 October 2015.
4 Crooks, Ed, and Hornby, Lucy, 'Sunshine Revolution: The Age of Solar Power', *Financial Times*, 5 November 2015.
5 BLS, Current Employment Statistics.
6 From the author's own conversations with Michael Mandel.
7 BLS, ibid.
8 'The Digital Degree', *The Economist*, 28 June 2014.
9 'Wealth by Degrees', ibid.
10 Ellison, Glenn, and Fisher Ellison, Sara, 'Match Quality, Search, and the Internet Market for Used Books', September 2014.
11 'Silver Lining', *The Economist*, 4 October 2014.
12 Hall, Jonathan, and Krueger, Alan, 'An Analysis of the Labor Market for Uber's Driver-Partners in the United States'.

4. The Virtues of Scarcity

1 Chatterji, Aaron, and Fairlie, Robert, 'High-technology Entrepreneurship in Silicon Valley', *Journal of Economics and Management Strategy*, Summer 2013.
2 Ibid.
3 BEA, Regional Economic Accounts.
4 S & P Case-Shiller Home Prices Indexes.
5 The relationships between economic booms, high housing costs and migration are given a more thorough treatment in my 2011 title, *The Gated City* (Kindle Single).
6 The Reverend Thomas Robert Malthus (1766–1834) was an English cleric and scholar, influential in the fields of political economy and demography. He was the author of *An Essay on the Principle of Population* (London: J. Johnson, 1798).
7 Federico, Giovanni, 'Growth, Specialisation and Organisation of World Agriculture', in Neal, Larry, and Williamson, Jeffrey G., eds., *Cambridge History of Capitalism* (Cambridge: Cambridge University Press, 2014).

8 Wright, Gavin, *Sharing the Prize: The Economics of the Civil Rights Revolution in the American South* (Cambridge, MA: Harvard University Press, 2013).

9 'A Proper Reckoning', *The Economist*, 12 March 2016.

10 But for the low wage growth resulting from immigration, New York's textile sector would have probably been much smaller and less labour-intensive, as jobs requiring more and cheaper labour moved elsewhere.

11 Nor are workers in source countries necessarily harmed. Emigration can facilitate the flow of ideas and money to the source country from destination countries. Many immigrants also eventually return, having gained valuable experience abroad. And the prospect of jobs abroad can encourage workers in source countries to invest more heavily in human capital – an investment which benefits the source country, since not all of those who receive training will, in fact, travel abroad.

12 OECD, Trade Union Density (http://stats.oecd.org/Index.aspx?Data SetCode=UN_DEN).

13 While private-sector unions have declined in political and economic importance, public-sector unions remain a force to be reckoned with, but they occupy an uncomfortable position within modern economies, as the group from which they attempt to extract additional surplus is the body of taxpayers. Political battles over the generosity of public-sector pay and pensions, for instance, in countries facing budget woes have earned public-sector unions the enmity of rich and poor alike.

14 Bengtsson, Erik, 'Labour's Share in Sweden, 1850–2000', *Abstract*, September 2012; Karabarbounis, Loukas, and Neiman, Brent, 'The Global Decline of the Labor Share', *Quarterly Journal of Economics*, June 2013.

15 The political ramifications of labour scarcity are not as intuitive as one might imagine. One might guess, for example, that in historical periods in which labour is relatively scarce, and in which workers therefore stand to wield substantial economic power, repressive societies would find themselves forced to acquiesce to worker grievances. In fact, much the opposite has often been the case. In Russia, for instance, repressive serfdom seemed to emerge as workers became increasingly scarce relative to available land, as a result of the migration of workers towards land newly conquered by the expanding Russian empire. When labour was cheap and abundant, there was little reason for the powerful to resort to the threat of state violence; willing workers could be obtained easily enough at market wages. It was when labour became scarce that the Russian political elite prevailed on the Tsar to grant them the

authority to tie workers ever closer to the land. As labour became scarce relative to land, landowners worked to take what was effectively an ownership stake in labour.

In other historical contexts, however, scarcity has been a source of power. In the late Middle Ages, for example, workers whose numbers had been greatly reduced by plague found it relatively easy to escape serfdom by moving to vacated land, leading to an erosion in the political order of feudalism. This erosion was sometimes accelerated by the inability of landowners to act collectively, as they competed against each other to retain labour. More landowners therefore found themselves acquiring labour services by offering contracts to the workers they needed, rather than by compelling them to live on and work their land.

When there is political competition among elites, labour scarcity boosts worker economic and political power. When elite interests are powerful and aligned, labour scarcity tends to result in repression.

16 Smith, *The Wealth of Nations*.
17 Acemoglu, Daron, and Robinson, James A., *Why Nations Fail: The Origins of Power, Prosperity, and Poverty* (London: Profile Books, 2012)

5. The Firm as an Information-Processing Organism

1 OECD, Entrepreneurship at a Glance 2015, August 2015.
2 Coase, R. H., 'The Nature of the Firm', *Economica*, Vol. 4, No. 16 (Nov. 1937).
3 Ocean Tomo, 'Annual Study of Intangible Asset Market Value', LLC, 2015.
4 Clayton M. Christensen (1952–), Kim B. Clark Professor of Business Administration at the Harvard Business School, and author of *The Innovator's Dilemma: When New Technologies Cause Great Firms to Fail* (Cambridge, MA: Harvard Business Review Press, 1997).
5 Lepore, Jill, 'The Disruption Machine', *New Yorker*, 23 June 2014.
6 For more detail on this process, see, for instance: Henderson, Rebecca, and Kaplan, Sarah, 'Inertia and Incentives: Bridging Organisational Economics and Organisational Theory', NBER Working Paper 11849, 2005.
7 'State of the News Media 2015', Pew Research Center, 29 April 2015.
8 Henderson, Rebecca, 'Investment and Incompetence as Responses to Radical Innovation: Evidence from the Photolithographic Alignment Equipment Industry', *RAND Journal of Economics*, Vol. 24, No. 2 (Summer, 1993).

9 Bresnahan, Timothy, Greenstein, Shane, and Henderson, Rebecca, 'Schumpeterian Competition and Diseconomies of Scope: Illustrations from the Histories of Microsoft and IBM', Harvard Business School Working Paper 11-077, January 2011.

10 Song, Jae, Price, David, Guvenen, Fatih, Bloom, Nicholas, and von Wachter, Till, 'Firming Up Inequality', NBER Working Paper 21199, May 2015.

6. Social Capital in the Twenty-First Century

1 'Bigmouth Strikes Again', *The Economist*, 14 September 2013.

2 Engels, Friedrich, *Die Lage der arbeitenden Klasse in England* (1845), which was translated into English by Mrs F. Kelley Wischnewetzky and published as *The Condition of the Working Class in England* (1887).

3 Putnam, Robert, 'Bowling Alone: America's Declining Social Capital', *Journal of Democracy*, January 1995. Robert David Putnam (1941–) is a political scientist and Malkin Professor of Public Policy at the Harvard University John F. Kennedy School of Government. His most famous publication is *Bowling Alone: The Collapse and Revival of American Community* (New York, NY: Simon & Schuster, 2001), developed from his 1995 essay of the same name.

4 A'hearn, Brian, 'The British Industrial Revolution in a European Mirror', Floud, Roderick, Humphries, Jane, and Johnson, Paul, eds., *The Cambridge Economic History of Modern Britain* (Cambridge: Cambridge University Press, 2014).

5 Allen, Robert, 'Engels' Pause: Technical Change, Capital Accumulation, and Inequality in the British Industrial Revolution', *Explorations in Economic History*, 8 February 2008.

6 Marx, Karl, and Engels, Friedrich, *Manifesto of the Communist Party* (1848).

7 Mokyr, Joel, 'The Rise and Fall of the Factory System: Technology, Firms, and Households Since the Industrial Revolution', Carnegie-Rochester Conference Series on Public Policy, December 2001.

8 Piketty, *Capital in the Twenty-First Century*.

9 Mokyr, 'The Rise and Fall of the Factory System'.

10 Ibid; the internal quote is from Mitch, David, 'The Role of Education and Skill in the British Industrial Revolution', Mokyr, Joel, ed. *The British Industrial Revolution: An Economic Perspective*, 1993.

11 Subramanian, Arvind, and Kessler, Martin, 'The Hyperglobalization of Trade and its Future', Word Trade Organization, 2013.

12 Dedrick, Jason, Kraemer, Kenneth, and Linden, Greg, 'Who Profits from Innovation in Global Value Chains? A Study of the iPod and Notebook PCs', *Industrial and Corporate Change*, February 2010.

13 Marshall, Alfred, *Principles of Economics* (1890).

14 Copeland, Rob, and Hope, Bradley, 'Schism atop Bridgewater, the World's Largest Hedge Fund', *Wall Street Journal*, 5 February 2016.

15 Robischon, Noah, 'How BuzzFeed's Jonah Peretti is Building a 100-year Media Company', *Fast Company*, 16 February 2016.

16 This sort of writing is business as usual at *The Economist*, but did represent a departure from the typical fare offered by other American news publications, operating on a daily rhythm.

17 Brookings Institution, Global Metro Monitor (http://www.brookings.edu/research/reports2/2015/01/22-global-metro-monitor).

18 Acemoglu, Daron, and Robinson, James, 'Why Did the West Extend the Franchise? Democracy, Inequality, and Growth in Historical Perspective', *Quarterly Journal of Economics*, November 2000.

7. Playgrounds of the 1 per cent

1 Cairncross, Frances, *The Death of Distance: How the Communications Revolution Is Changing Our Lives* (Cambridge, MA: Harvard Business School Press, 1997).

2 UK Office for National Statistics.

3 US Census Bureau

4 Glaeser, Edward, and Resseger, Matthew, 'The Complementarity Between Cities and Skills', *Journal of Regional Science*, 13 January 2010.

5 Berger, Thor, and Frey, Carl Benedikt, 'Technology Shocks and Urban Evolutions: Did the Computer Revolution Shift the Fortunes of US Cities?', June 2014.

6 Berry, Christopher, and Glaeser, Edward, 'The Divergence of Human Capital Levels Across Cities', *Papers in Regional Science*, 2005.

7 Robert-Nicoud, Frédéric and Hilber, Christian, 'On the Causes and Consequences of Land Use Regulations', VoxEU, 18 March 2013.

8 'Gotham on Thames', *The Economist*, 27 February 2016.

9 Avent, Ryan, *The Gated City*.

10 Hsieh, Chang-Tai, and Moretti, Enrico, 'Why Do Cities Matter? Local Growth and Aggregate Growth', NBER Working Paper, May 2015.

11 Ganong, Peter, and Shoag, Daniel, 'Why Has Regional Income Convergence in the US Declined?', Harvard Kennedy School Working Paper, March 2013.

12 Rognlie, Matthew, 'Deciphering the fall and rise in the net capital share', Brookings Papers on Economic Activity, January 2015; 'The Paradox of Soil', *The Economist*, 2015.

13 Schleicher, David, 'City Unplanning', *Yale Law Journal*, May 2013.

8. Hyperglobalization and the Never-Developing World

1 Pritchett, Lant, 'Divergence, Big Time', *Journal of Economic Perspectives*, Summer 1997.

2 Milanovic, Branko, 'Global Inequality by the Numbers: In History and Now, an Overview', The World Bank, 2012.

3 IMF, Word Economic Outlook Database (http://www.imf.org/external/pubs/ft/weo/2016/01weodata/index.aspx).

4 Maddison Project Database.

5 'The Headwinds Return', *The Economist*, 13 September 2014.

6 The World Bank; Center for Global Development.

7 Robert Lucas Jr (1937–) is an American economist at the University of Chicago. In 1995 he won the Nobel Prize in Economic Sciences.

8 Lucas, Robert E., Jr, 'On the Mechanics of Economic Development', *Journal of Monetary Economics*, 1988.

9 Subramanian and Kessler, 'The Hyperglobalization of Trade and its Future'.

10 Terence James 'Jim' O'Neill, Baron O'Neill of Gatley (1957–), is a British economist and the former chairman of Goldman Sachs Asset Management. He is best known for coining BRIC, the acronym that stands for Brazil, Russia, India and China – the four rapidly developing countries that have come to symbolize the shift in global economic power away from the developed G7 economies.

11 Baldwin, Richard, 'Trade and Industrialization after Globalization's Second Unbundling: How Building and Joining a Supply Chain are Different and Why it Matters' in Feenstra, Robert C., and Taylor, Alan M., eds., *Globalization in an Age of Crisis: Multilateral Economic Cooperation in the Twenty-First Century* (Chicago, IL: University of Chicago Press, 2014).

12 Subramanian, Arvind, 'Premature De-industrialization', Center for Global Development, 22 April 2014.

13 'Arrested Development', *The Economist*, 4 October 2014.
14 Amirapu, Amrit, and Subramanian, Arvind, 'Manufacturing or Services? An Indian Illustration of a Development Dilemma', Center for Global Development Working Paper 409, June 2015.
15 Ibid.
16 Pritchett, Lant, and Summers, Larry, 'Asiaphoria Meets Regression to the Mean', NBER Working Paper 20573, October 2014.

9. The Scourge of Secular Stagnation

1 Hansen, Alan Harvey, *Full Recovery or Stagnation* (New York, NY: W. W. Norton & Company, 1938).
2 Ben Shalom Bernanke (1953–) is an American economist who served two terms as chairman of the Federal Reserve, the central bank of the United States, from 2006 to 2014. During his tenure as chairman, Bernanke oversaw the Federal Reserve's response to the late-2000s financial crisis.
3 Bernanke, Ben, in his speech 'The Global Saving Glut and the US Current Account Deficit', 10 March 2005.
4 Bank for International Settlements, Total Credit to Households as a Percentage of GDP (http://stats.bis.org/statx/srs/table/f3.1).
5 Federal Reserve Bank of New York, Household Debit and Credit Report.
6 Wolff, Edward, 'Household Wealth Trends in the United States, 1962–2013: What Happened Over the Great Recession?', NBER Working Paper 20733, December 2014.

10. Why Higher Wages are so Economically Elusive

1 For more on how such a world might work, see Saadia, Manu, *Trekonomics: The Economics of Star Trek* (San Francisco, CA: Pipertext, 2016).
2 There is a line of argument that suggests that, in the absence of a rising minimum wage, wage subsidies simply spare firms the need to ever give employees a rise. Studies suggest that this is generally not the case; the workers capture most of the benefit of the wage subsidy rather than the firm. See 'Credit Where Taxes are Due', *The Economist*, 2015.
3 BLS.
4 Economist Anthony Atkinson, for instance, proposes a citizen's income or participant's income, for which voluntary service work would qual-

ify an individual just as market work would. See Atkinson, Anthony B., *Inequality: What Can Be Done?* (Cambridge, MA: Harvard University Press, 2015).

5 Such a story had not, to my knowledge, been told at the time of writing.

6 'The Gifts of the Moguls', *The Economist*, 4 July 2015.

11. The Politics of Labour Abundance

1 Yoshihiro Francis Fukuyama (1952–) is an American political scientist, political economist and author, known for his book *The End of History and the Last Man* (New York, NY: Free Press, 1992), which expanded on his 1989 essay, 'The End of History'.

2 Schleicher, David, 'Things Aren't Going That Well Over There Either: Party Polarization and Election Law in Comparative Perspective', University of Chicago Legal Forum, 18 November 2014.

3 US Census Bureau, Income and Poverty.

4 Kenworthy, Lane, and Pontusson, Jonas, 'Rising Inequality and the Politics of Redistribution in Affluent Countries', *Perspectives on Politics*, September 2005.

5 Schleicher, 'Things Aren't Going That Well Over There Either'.

6 Aguiar, Mark, Hurst, Erik, and Karabarbounis, Loukas, 'Time Use During the Great Recession', *American Economic Review*, 5 August 2013.

7 'Goldilocks Nationalism', *The Economist*, 27 September 2014.

8 Chait, Jonathan, 'Will the Supreme Court Just Disappear?', *New York Magazine*, 21 February 2016.

9 Reportedly first said in a 1904 speech, and quoted by the IRS above the entrance to their headquarters at 1111 Constitution Avenue, Washington, DC.

12. Human Wealth

1 Smith, *The Wealth of Nations*.

2 Smith, Adam, *The Theory of Moral Sentiments* (London: A. Millar, 1759).

Further Reading

Acemoglu, Daron, and Robinson, James A., *Why Nations Fail: The Origins of Power, Prosperity, and Poverty* (London: Profile Books, 2012)

Atkinson, Anthony B., *Inequality: What Can Be Done?* (Cambridge, MA: Harvard University Press, 2015)

Boustan, Leah Platt, Frydman, Carola, and Margo, Robert A., eds., *Human Capital in History: The American Record* (Chicago, IL: University of Chicago Press, 2014)

Brynjolfsson, Erik, and McAfee, Andrew, *Race Against the Machine: How the Digital Revolution is Accelerating Innovation, Driving Productivity, and Irreversibly Transforming Employment and the Economy* (Digital Frontier Press, 2011)

——, *The Second Machine Age: Work, Progress, and Prosperity in a Time of Brilliant Technologies* (New York, NY: W. W. Norton & Company, 2014)

Cairncross, Frances, *The Death of Distance: How the Communications Revolution is Changing Our Lives* (Cambridge, MA: Harvard Business School Press, 1997)

Christensen, Clayton M., *The Innovator's Dilemma: When New Technologies Cause Great Firms to Fail* (Cambridge, MA: Harvard Business Review Press, 1997)

Cowen, Tyler, *Average is Over: Powering America Beyond the Age of the Great Stagnation* (New York, NY: E. P. Dutton & Co Inc., 2013)

——, *The Great Stagnation: How America Ate All the Low-Hanging Fruit of Modern History, Got Sick, and Will (Eventually) Feel Better* (New York, NY: E. P. Dutton & Co Inc., 2011)

Edgerton, David, *The Shock of the Old: Technology and Global History since 1900* (London: Profile Books, 2006)

Engels, Friedrich, *The Condition of the Working Class in England* (1887).

Feenstra, Robert C., and Taylor, Alan M., eds., *Globalization in an Age of Crisis: Multilateral Economic Cooperation in the Twenty-First Century* (Chicago, IL: University of Chicago Press, 2014)

Floud, Roderick, Humphries, Jane, and Johnson, Paul, eds., *The Cambridge Economic History of Modern Britain* (Cambridge: Cambridge University Press, 2014)

Ford, Martin, *The Lights in the Tunnel: Automation, Accelerating Technology and the Economy of the Future* (Createspace, 2009)

———, *Rise of the Robots: Technology and the Threat of a Jobless Future* (London: Oneworld Publications, 2015)

Friedman, Milton, and Schwartz, Anna, *A Monetary History of the United States, 1867–1960* (Princeton, NJ: Princeton University Press, 1963)

Fukuyama, Francis, *The End of History and the Last Man* (The Free Press, 1992)

Glaeser, Edward, *Triumph of the City: How Our Greatest Invention Makes Us Richer, Smarter, Greener, Healthier, and Happier* (London: Macmillan, 2011)

Goldin, Claudia and Katz, Lawrence, *The Race Between Education and Technology* (Cambridge, MA: Harvard University Press, 2008)

Gordon, Robert, *The Rise and Fall of American Growth: The U.S. Standard of Living since the Civil War* (Princeton, NJ: Princeton University Press, 2016)

Hansen, Alan Harvey, *Full Recovery or Stagnation* (New York, NY: W. W. Norton & Company, 1938)

Hayes, Christopher, *Twilight of the Elites: America After Meritocracy* (New York, NY: Crown Publishing Group, 2012)

Keynes, John Maynard, *Essays in Persuasion*, John Maynard Keynes (London: Macmillan, 1931)

Landau, Ralph, Taylor, Timothy, and Wright, Gavin, eds., *The Mosaic of Economic Growth* (Palo Alto, CA: Stanford University Press, 1995)

Larson, Erik, *The Devil in the White City: Murder, Magic and Madness at the Fair that Changed America* (New York, NY: Crown Publishing Group, 2003)

Mason, Paul, *Postcapitalism: A Guide to Our Future* (London: Allen Lane, 2015)

Malthus, Thomas, *An Essay on the Principle of Population* (London: J. Johnson, 1798)

Marx, Karl, and Engels, Friedrich, *Manifesto of the Communist Party* (1848)

Milanovic, Branko, *Global Inequality: A New Approach for the Age of Globalization* (Cambridge, MA: Harvard University Press, 2016)

Mokyr, Joel, *The Gifts of Athena: Historical Origins of the Knowledge Economy* (Princeton, NJ: Princeton University Press, 2002)

———, *The Lever of Riches: Technological Creativity and Economic Progress* (Oxford: Oxford University Press, 1990)

Moretti, Enrico, *The New Geography of Jobs* (New York, NY: Houghton Mifflin Harcourt, 2012)

Murray, Charles, *Coming Apart: The State of White America, 1960–2010* (New York, NY: Crown Publishing Group, 2012)

Pickett, Kate, and Wilkinson, Richard, *The Spirit Level: Why Greater Equality Makes Societies Stronger* (London: Allen Lane, 2009)

Piketty, Thomas, *Capital in the Twenty-First Century* (Cambridge, MA: Harvard University Press, 2014)

Putnam, Robert, *Bowling Alone: The Collapse and Revival of American Community* (New York, NY: Simon & Schuster, 2001)

Rifkin, Jeremy, *The Zero Marginal Cost Society: The Internet of Things, the Collaborative Commons, and the Eclipse of Capitalism* (London: Palgrave Macmillan, 2014)

Rodrik, Dani, *The Globalization Paradox: Democracy and the Future of the World Economy* (Oxford: Oxford University Press, 2011)

Saadia, Manu, *Trekonomics: The Economics of Star Trek* (San Francisco, CA: Pipertext, 2016)

Shirky, Clay, *Cognitive Surplus: Creativity and Generosity in a Connected Age* (London: Allen Lane, 2010)

Smith, Adam, *An Inquiry into the Nature and Causes of the Wealth of Nations* (London: W. Strahan and T. Cadell, 1776)

———, *The Theory of Moral Sentiments* (London: A. Millar, 1759)

Weil, David, *The Fissured Workplace: Why Work Became So Bad and What Can be Done to Improve It* (Cambridge, MA: Harvard University Press, 2014)

Wright, Gavin, *Sharing the Prize: The Economics of the Civil Rights Revolution in the American South* (Cambridge, MA: Harvard University Press, 2013)

Acknowledgements

This book is the fruit of nearly ten years spent covering and thinking about the world economy, and of two years of writing and editing. Looking back, I am struck by how many people have had a hand in the project and the forming of the ideas within it. To a great extent, the book is simply the latest comment in a running conversation kept up by scores of economists, journalists and other writers, whose work I have followed closely over the last decade and cited in the text: my two cents atop a wealth of analysis and argument.

Many others played a more personal role in shaping this contribution. I am indebted to many mentors and colleagues at *The Economist*. Thanks to Andrew Palmer and Ed McBride, masterful editors whose guidance has made me a much better writer and journalist. Many of the ideas in the book began as briefings, put together with the editing and oversight of Oliver Morton, who repeatedly helped me find the core idea I had been struggling to bring out. I am similarly grateful to Barbara Beck, for her help on the special report that became the basis for the book. Many of my views on these matters were forged during our editorial debates, and shaped by the insights of too many amazing colleagues to name. I will single out, however, Tom Standage, Emma Duncan, Ludwig Siegele and Tim Cross, discussions with whom were particularly informative. Most importantly, I am indebted to Zanny Minton Beddoes, without whose confidence and trust I would not have found myself in this position, and whose brilliance has made me a better thinker and writer.

The ideas in the book were also shaped by years of debate and discussion with fellow economics writers and bloggers. I am especially grateful to Tyler Cowen, Matthew Yglesias, Karl Smith, Steve Randy

Waldman and Brad DeLong, whose blogs have been a trusted sound-board off which I could bounce ideas.

The book itself has been moulded by many hands. The text was immeasurably improved thanks to comments on early drafts from David Schleicher and Soumaya Keynes, and it was a great pleasure to work with Anna Hervé, who helped shape and polish the text. I am thankful for the patience and insight of my editors, Tom Penn and Tim Bartlett, who spent long hours helping me find the book in my wandering manuscript, and this book would not exist at all without the steady, guiding hand of my literary agent, Jonathan Conway, who has never failed to provide a desperately needed word of encouragement or suggestion.

Above all, I am endlessly grateful to my wife, who was there across the whole of the long slog, fighting the battles with me, commiserating when the going threatened to become too tough to continue, contributing in too many ways to count. Thank you, Lisa, for your love, inspiration and support. I wouldn't have made it without you.

Index